Guide for the Total Solar Eclipse of 2030

Color Edition

Fred Espenak

Edition 1.0
November 2024

Guide for the Total Solar Eclipse of 2030 – Color Edition

Astropixels Publishing
P.O. Box 16197
Portal, AZ 85632

Astropixels Publishing Website: *www.astropixels.com/pubs*

This book may be ordered at: *www.astropixels.com/pubs/Guide2030.html*

Printed in the United States of America

ISBN 978-1-941983-50-8

Astropixels Publication: AP049 (Version 1.0b)

First Edition

Front Cover: Maps of the path of the 2030 total solar eclipse through Africa and Australia.

Back Cover: Fred Espenak in Libya during the total solar eclipse of 2006 ©2006, Patricia Espenak.

More about the eclipse can be found at:

www.eclipsewise.com/solar/SEnews/TSE2030.html

Table of Contents

Photo 1 – Two cameras were used to create this dramtic view of the 2001 total eclipse in Zambia. One camera captured totality and a thorn acacia tree while the second camera recorded the partial phases at 5 minute intervals through a solar filter. Total Solar Eclipse of June 21, 2001, Chisamba, Zambia (©2001 Fred Espenak, www.MrEclipse.com)

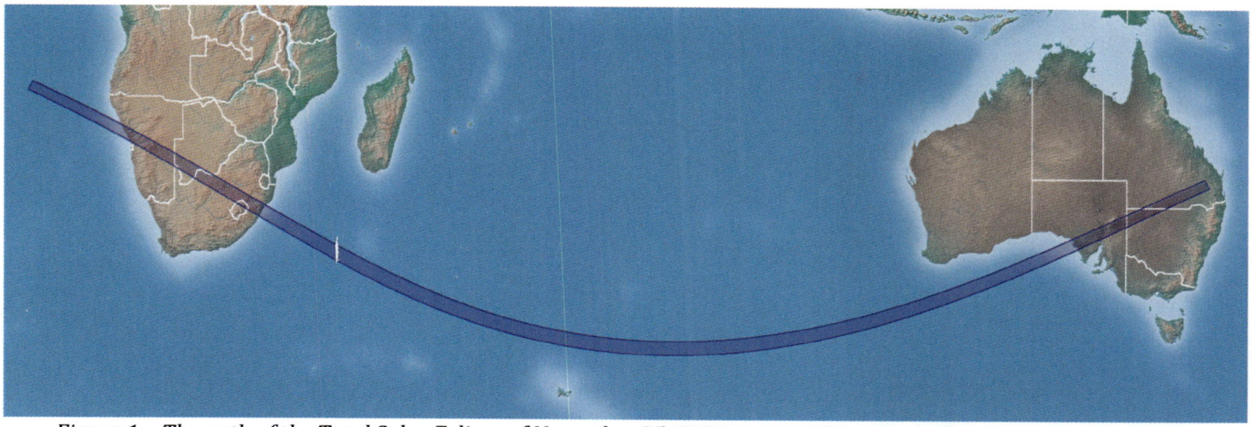

Figure 1 – The path of the Total Solar Eclipse of November 25, 2030 crosses Africa, the Indian Ocean, and Australia.

1 – Introduction

On Monday, November 25, 2030, a total eclipse of the Sun will be visible from the Eastern Hemisphere. Although a partial eclipse will be seen from southern Africa, the Indian Ocean, Antarctica, and Australia, the total phase in which the Moon completely covers the Sun (known as totality) will only be seen from within the narrow path of the Moon's umbral shadow as it sweeps across parts of Africa and Australia (Figure 1).

During totality, the Sun's faint and exquisitely beautiful outer atmosphere – the solar corona – can be seen with the naked eye, the landscape is plunged into an eerie twilight, and brighter stars and planets appear. A total solar eclipse is one of the most awesome natural spectacles, but to witness it you must be in the *path of totality*.

The path of the Moon's umbral shadow begins in the Atlantic Ocean and travels east to reach the coast of southern Africa where it crosses parts of Namibia, Botswana, South Africa, and Lesotho. The central line duration of totality across southern Africa ranges from 1 minute 32 seconds to 2 minutes 43 seconds. Continuing across the Indian Ocean, greatest eclipse occurs at sea with a duration of 3 minutes 43.5 seconds.

The path slowly curves to the northeast where it once again intercepts land in southeastern Australia. Parts of the states of South Australia (SA), New South Wales (NSW), and Queensland (QLD) all lie within the path of totality. The eclipse track ends in Queensland (QLD) about 225 km (140 mi) northwest of Brisbane. The duration of totality near the sunset point is 1 minute 32 seconds.

A partial eclipse will be visible from a much larger region including southern Africa, Australia, New Zealand, southern Indonesia, Papua New Guinea, and Antarctica. Eclipse glasses or some type of eye protection must be used for viewing the partial eclipse. Of course, those in the path of the total eclipse must remove protective glasses to witness totality itself.

This guide is designed to assist observers in locating and reaching the central eclipse path to enjoy the awesome experience of totality.

Photo 2 – The 2017 corona is shown with the Diamond Ring before and after totality during the total solar eclipse of August 21, 2017, Casper, Wyoming. (©2017 Fred Espenak, www.MrEclipse.com)

Total Solar Eclipse of 2030 Nov 25

Greatest Eclipse = 06:51:36.9 TD (= 06:50:27.7 UTC)

Eclipse Magnitude = 1.0468 Saros Series = 133

Gamma = -0.3867 Saros Member = 46 of 72

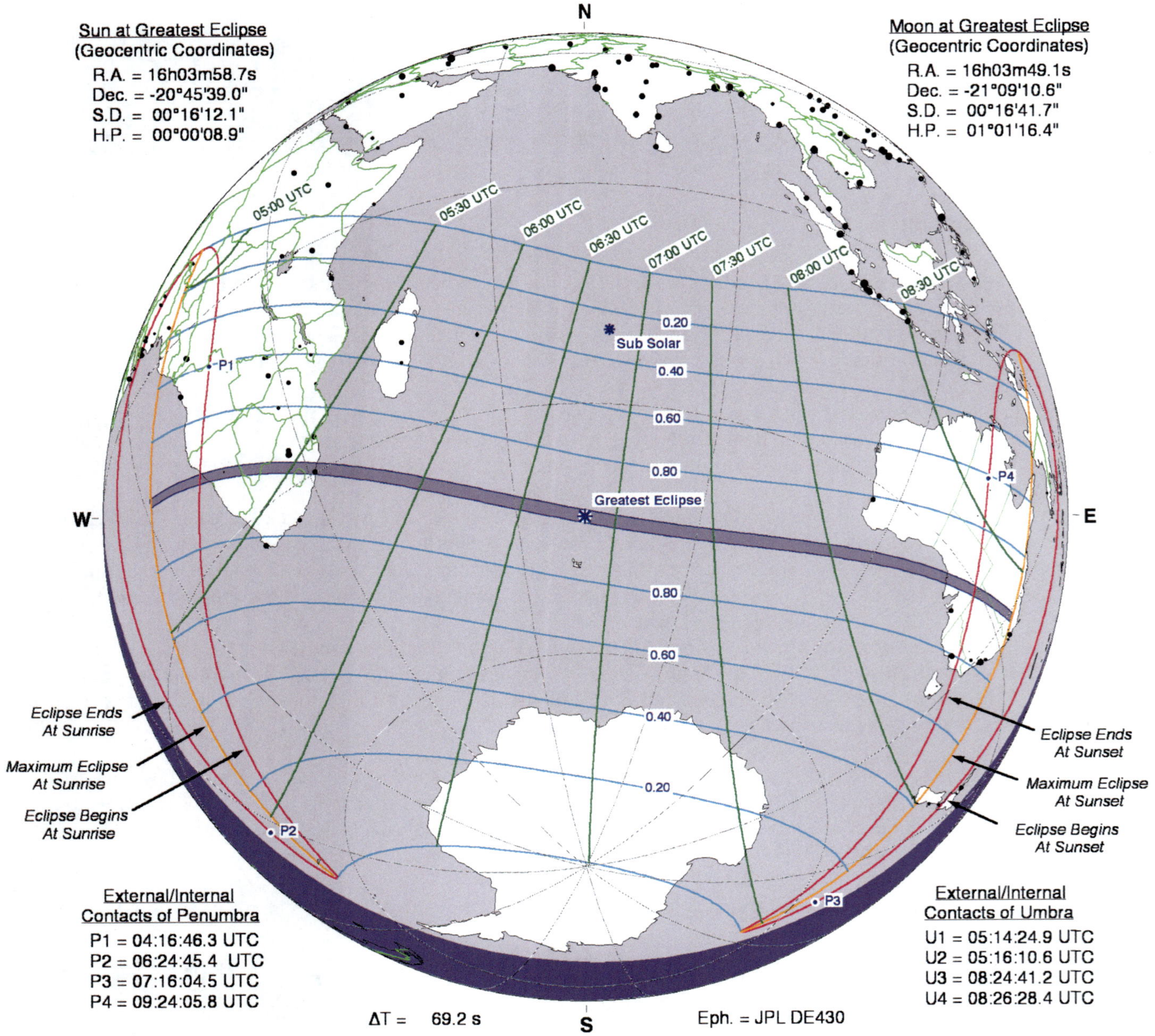

External/Internal Contacts of Penumbra

P1 = 04:16:46.3 UTC
P2 = 06:24:45.4 UTC
P3 = 07:16:04.5 UTC
P4 = 09:24:05.8 UTC

External/Internal Contacts of Umbra

U1 = 05:14:24.9 UTC
U2 = 05:16:10.6 UTC
U3 = 08:24:41.2 UTC
U4 = 08:26:28.4 UTC

ΔT = 69.2 s Eph. = JPL DE430

Circumstances at Greatest Eclipse: 06:50:27.7 UTC

Lat. = 43°36.6'S	Sun Alt. = 67.0°
Long. = 071°12.4'E	Sun Azm. = 7.0°
Path Width = 169.3 km	Duration = 03m43.53s

Circumstances at Greatest Duration: 06:52:00.6 UTC

Lat. = 43°42.7'S	Sun Alt. = 67.0°
Long. = 071°59.8'E	Sun Azm. = 4.1°
Path Width = 169.2 km	Duration = 03m43.56s

0 1000 2000 3000 4000 5000
Kilometers

Figure 2 – Orthographic projection map of the Total Solar Eclipse of 2030 Nov 25. Based on map from ***21st Century Canon of Solar Eclipses*** *by Fred Espenak (2017)*

2 – Orthographic Projection Map of the 2030 Total Solar Eclipse Path

The orthographic projection map of Earth in *Figure 2* shows the path of penumbral (partial eclipse) and umbral (total eclipse) shadows of the Moon during the eclipse.

The limits of the Moon's penumbral shadow define the region of visibility of the partial eclipse. It covers about half of Earth's daylight hemisphere and consists of several distinct zones or limits. At the northern and southern boundaries lie the limits of the penumbra's path. Great loops at the western and eastern extremes of the penumbra's path identify the areas where the eclipse begins and ends at sunrise and sunset, respectively.

Bisecting the *eclipse begins and ends at sunrise and sunset* loops is the curve of maximum eclipse at sunrise (western loop) and sunset (eastern loop). Points *P1* and *P4* mark the coordinates where the penumbral shadow first contacts (partial eclipse begins) and last contacts (partial eclipse ends) Earth's surface. The path of the umbral shadow (*path of totality*) appears within the penumbral path. Curves of constant eclipse magnitude define points where the maximum magnitude has values of 0.2, 0.4, 0.6, and 0.8.

Above the map is the instant of greatest eclipse[1], in Terrestrial Time[2] (TT) and Coordinated Universal Time[3] (UTC). The *Eclipse Magnitude* is the geocentric ratio of diameters of the Moon and the Sun. *Gamma* is the minimum distance of the Moon's shadow axis from Earth's center in Earth radii at greatest eclipse. Finally, the *Saros Series* of the eclipse is listed, followed by the *Saros Member* that gives the sequence position of the eclipse in the Saros, and the total number of eclipses in the series

In the upper left and right corners are the geocentric coordinates of the Sun and the Moon, respectively, at the instant of greatest eclipse. They are:

R.A. - Right Ascension
Dec. - Declination
S.D. - Apparent Semi-Diameter
H.P. - Horizontal Parallax

To the lower left are the contact times of the Moon's penumbral shadow with Earth. They mark the times when the partial eclipse begins and ends, globally. To the lower right are the contact times of the Moon's umbral shadow with Earth. They mark the times when the total eclipse path begins and ends, globally.

At bottom left are the geographic coordinates of the position of greatest eclipse along with the local circumstances at that location (i.e., Sun altitude, Sun azimuth, path width and duration of totality).

At bottom right are the corresponding values for the position of greatest duration[4]. Although the geographic locations of greatest duration and greatest eclipse differ by about 66 km (41 mi) along the eclipse track, the difference in the duration of totality is only 0.03 second.

3 – Overview Maps of the Path of Totality

The overview maps in *Figure 3* and *Figure 4* show the path of the total eclipse through Africa and Australia, respectively. The white lines running across the eclipse path mark the position of mid-eclipse at 5-minute intervals.

The Mercator projection used in this map preserves local directions and shapes at the expense of inflating the size of objects the further they are from the equator.

---

[1] The instant of greatest eclipse occurs when the Moon's shadow axis passes closest to Earth's center.

[2] Terrestrial Time or TT is a time scale established by astronomers to calculate the location or planetary position of other objects as they relate to the center of Earth and its rotation.

[3] Coordinated Universal Time or UTC is the primary time standard that regulates world clocks and time. It is based on International Atomic Time (TAI) with leap seconds added at irregular intervals to compensate for the slowing of Earth's rotation. UTC is not adjusted for daylight saving time. It is effectively a successor to Greenwich Mean Time (GMT).

[4] Greatest duration is the time (and location) when the length of totality reaches its maximum value.

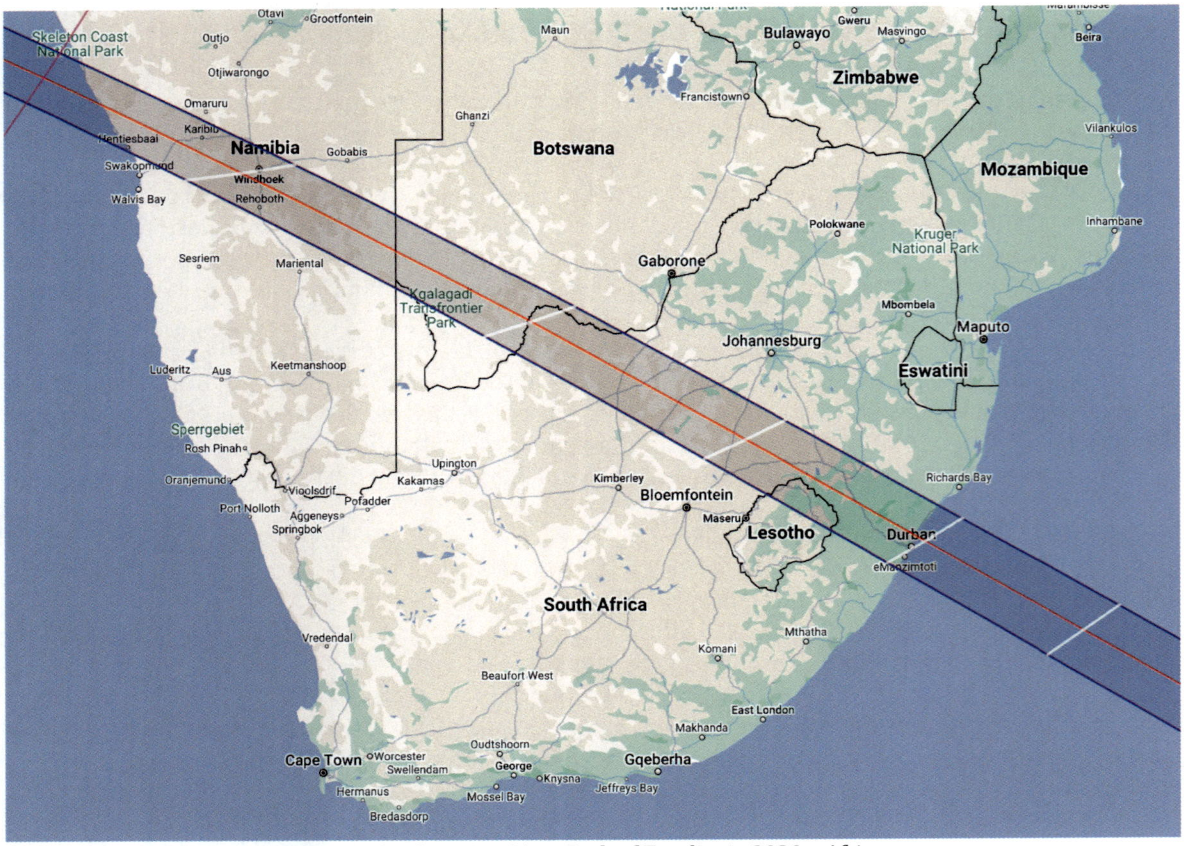

Figure 3 – Overview Map: Path of Totality in 2030 – Africa

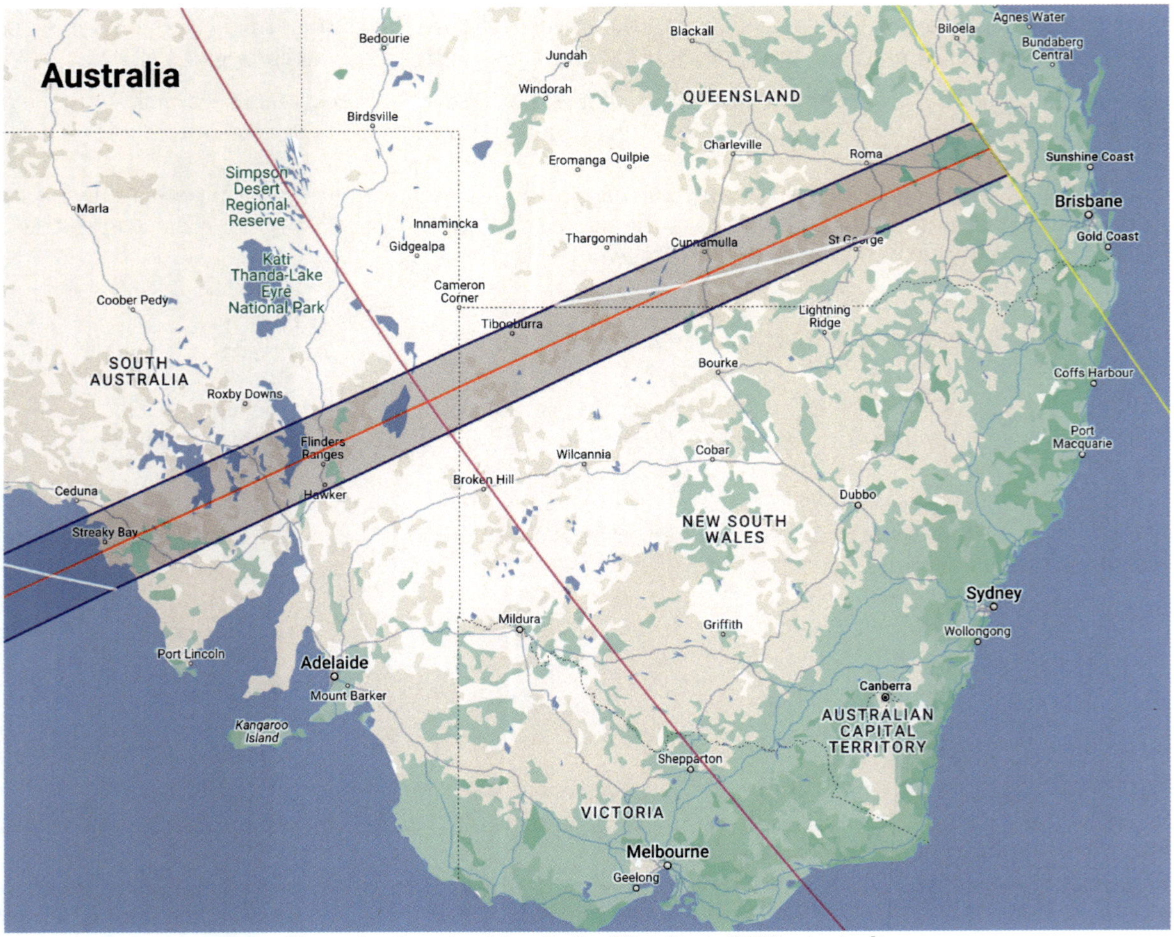

Figure 4 – Overview Map: Path of Totality in 2030 – Australia

Table 1 – Time Zones used by Countries within the 2030 Path of Totality

Time Zone Name	Abbr.	Offset	Locations in Path of Totality
Central Africa Time	CAT	UTC + 2 hrs	Namibia, Botswana
South Africa Standard Time	SAST	UTC + 2 hrs	South Africa, Lesotho
Australian Central Daylight Time	ACDT	UTC + 10.5 hrs	South Australia + Broken Hill, NSW
Australian Eastern Daylight Time	AEDT	UTC + 11 hrs	New South Wales
Australia Eastern Standard Time	AEST	UTC + 10 hrs	Queensland

4 – Path of Totality

The path of the Moon's umbral shadow begins at 05:15 UTC in the South Atlantic Ocean about 1400 km (870 mi) northwest of Namibia. Along the sunrise terminator, the duration of totality is 1m 30s (i.e., 1 minute 30 seconds) as seen from the center of the narrow, 117 km (73 mi) wide path.

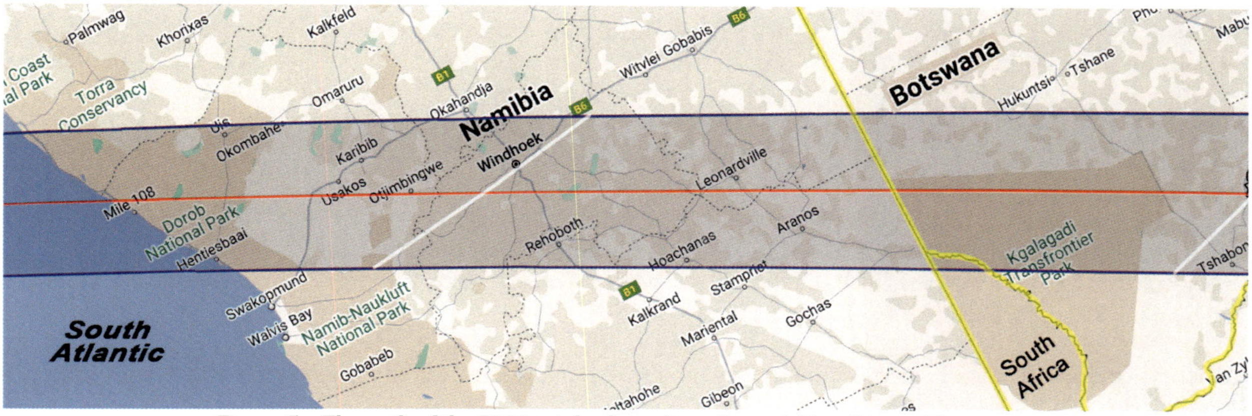

Figure 5 – The path of the 2030 total solar eclipse through Namibia and Botswana.

The umbra makes first landfall along Namibia's Skeleton Coast near Mile 108 on the C34 Highway in the Dorob National Park (Figure 5 and Map 01) at 05:18 UTC (07:18 am CAT). The central duration of totality is 1m 53s, the Sun's altitude is 13° and the track width is 133 km (83 mi). Dorob Park is known for its spectacular coastal dune belt and is Namibia's richest coastal area for birds.

Traveling inland with a velocity of 3.953 km/s (8840 mph), the shadow races across the Namibia's red deserts containing some of the largest sand dunes on Earth. Windhoek, Namibia's capital city (pop. 486,000), lies in the path of totality and is 26 km (16 mi) north of the central line. The duration of totality in the capital is 1m 52s and the Sun's altitude is 17°.

Continuing along a southeastern course, the path crosses several Namibian communities (Table 2, p. 12) before reaching the western border of Botswana at 05:22 UTC (07:22 am CAT). The central duration of totality is 2m 06s and the Sun's altitude is 20°.

In Botswana, much of the eclipse track crosses the Kgalagadi Transfrontier Park (Gemsbok National Park), a large wildlife preserve and conservation area. The park straddles South Africa and Botswana and is bisected by the Molopo River which forms Botswana's southern border.

Kgalagadi Transfrontier Park is located largely within the southern Kalahari Desert with a terrain of red dunes, sparse vegetation, and the dry riverbeds of the Nossob and Auob Rivers. The rivers are said to flow only about once per century. While prospects for clear sky on eclipse day are excellent, this region is for the adventuresome traveler who is an experienced camper.

The eclipse path crosses into South Africa along the Molopo River at 05:25 UTC (07:25 am CAT). The central duration of totality is 2m 13s and the Sun's altitude is 24°. South Africa offers many towns and villages from which to view totality, especially compared to Namibia and Botswana. This convenience occurs at the expense of increasing average cloud amount traveling east.

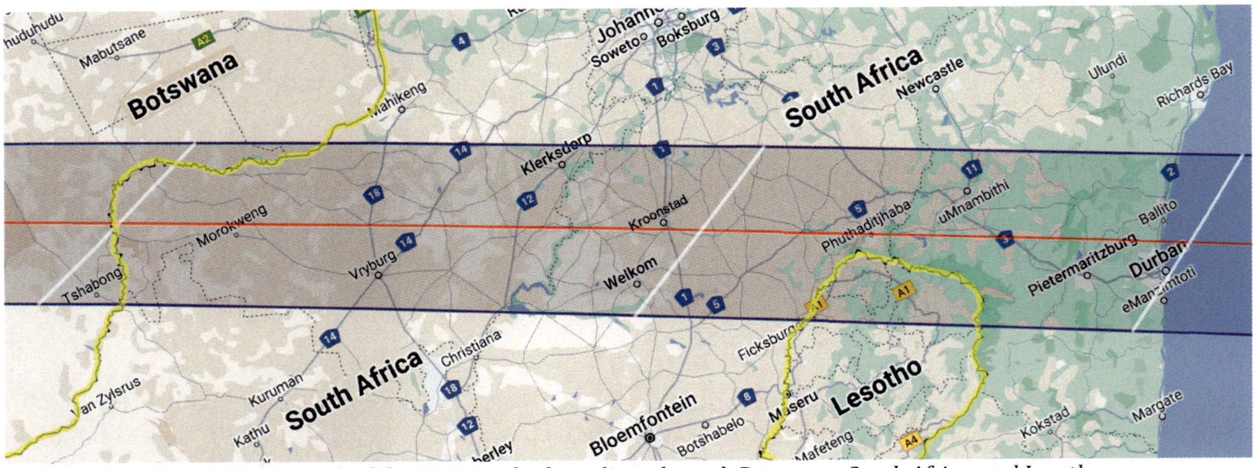

Figure 6 – The path of the 2030 total solar eclipse through Botswana, South Africa, and Lesotho.

Klerksdorp (pop. 275,000), one of the larger cities inside the path is located near the northern limit but still experiences a totality lasting 1m 52s (Figure 6). It's located 156 km (100 mi) southwest of Johannesburg making it easy to reach. One more popular destination for eclipse chasers will be Kroonstad (pop. 168,000). It's closer to the central line with 2m 24s of totality.

Another large city in the path is Pieteraritzburg (pop. 265,000) with a duration of 2m14s. It is most easily accessed from Durban (pop. 1,740,000), the largest city in the entire path of the 2030 total eclipse. For the citizens of Durban, mid-eclipse is at 15:35 UTC (07:35 am CAT) with a duration of 2m 18s. The central line passes 26 km (16 mi) north of Durban where totality lasts 2m 34s. Unfortunately, the average cloud amount in Durban is 67% compared to 34% in Klerksdorp (Table 8).

Northern Lesotho is also visited by the umbral shadow, but the central line misses the nation and remains in South Africa.

Leaving the African continent, the Moon's umbra spends the next 3 ¾ hours traversing over 9,000 km (5600 mi) of the Indian Ocean without crossing any land. Greatest eclipse occurs at 06:50:29 UTC with a central duration on 3m 43s and a Sun altitude of 67°. The nearest land to greatest eclipse is the remote Kerguelen Islands 650 km (400 mi) to the south.

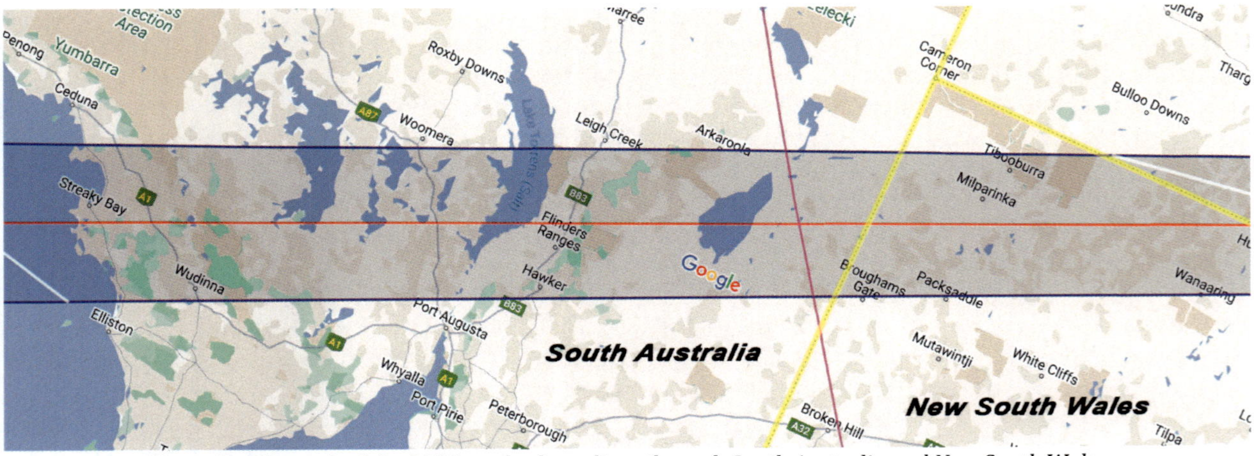

Figure 7 – The path of the 2030 total solar eclipse through South Australia and New South Wales.

The path of totality gradually curves to the northeast and reaches the southern coast of the Australian continent in the state of South Australia at 08:20 UTC (6:50 pm ACDT). The central line duration is 2m 02s and the Sun's altitude is 17°. (Figure 7 and Map 09).

Continuing inland, the path crosses Australia's Great Ocean Road (B100), an outstanding coastal highway with awe-inspiring limestone cliffs and incredible rock formations (in Victoria). Nearby is the A1, the longest national highway in the world, giving eclipse observers easy access to the path from Adelaide 600 km (370 mi) to the southeast via the A1.

The central line crosses Lake Torrens, a usually a dry salt flat with patches of seasonal water. Thunderstorms occasionally provide a small amount of water that attracts a variety of birdlife. Just east of Lake Torrens is B83, the Outback Highway that cuts across the path. The central line duration along the B83 is 1m 53s and the Sun's altitude is 13°.

Lake Frome, another large salt lake lies in the eclipse track just before the shadow leaves South Australia. The path of totality crosses the northwest corner of New South Wales (NSW), about 200 km (125 mi) north of Broken Hill. The shadow traverses NSW in just one minute before entering Queensland at 08:25 UTC (6:25 pm AEST). The central duration is 1m 43s and the Sun is 7° above the horizon.

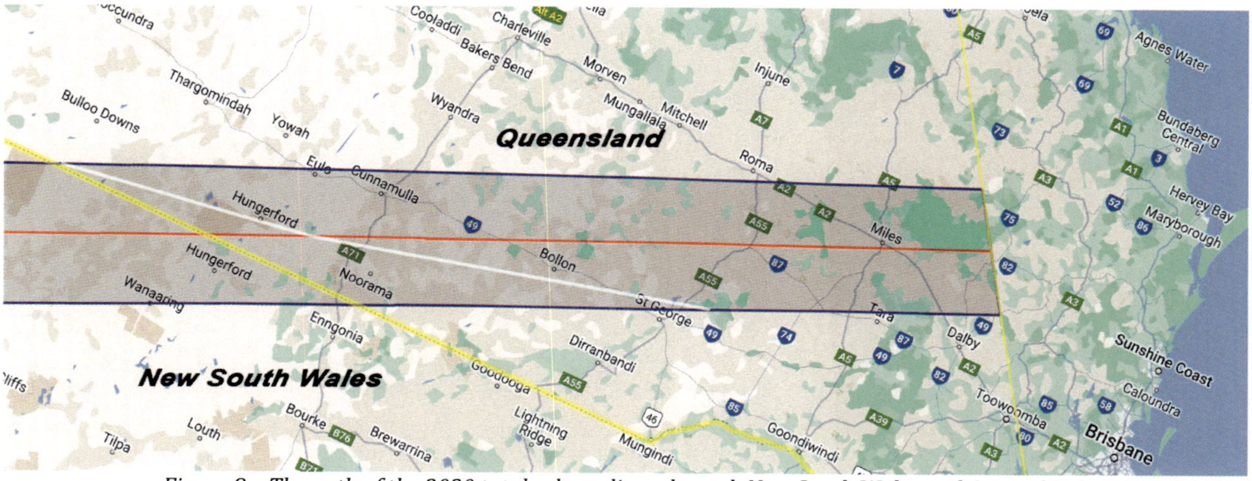

Figure 8 – The path of the 2030 total solar eclipse through New South Wales and Queensland.

It takes about 45 seconds for the axis of the umbra to cross Queensland where the path of totality ends 230 km (144 mi) northwest of Brisbane (pop. 2,536,000). The third largest city in Australia will see a magnitude 0.967 partial eclipse at sunset (6:24 AEST).

The umbra leaves Earth's surface at 08:26:28.4 UTC (Figure 8 and Map17). Over a period of 3 hours 10 minutes, the umbra travels along a 14,540 km (9,035 mi) long track that covers a region equal to 0.43% of Earth's surface area.

Photo 3 – The most recent total solar eclipse visible from Australia was on April 8, 2023.
*Shot aboard the **Pacific Explorer** off Western Australia's North West Cape, © 2023 F. Espenak, www.MrEclipse.com*

Table 2 – The 2030 Path of Totality

UTC	Latitude	Longitude	Diameter Ratio	Sun Alt	Sun Azm	Path Width km	Major Axis km	Minor Axis km	Duration Totality
05:15.3	16°12.8'S	001°34.5'E	1.030	0.0°	111.6°	116.6	-	100.3	01m30.9s
05:20	22°39.5'S	016°43.0'E	1.035	16.6°	106.0°	137.7	411.3	117.4	01m59.5s
05:30	27°51.5'S	027°29.3'E	1.039	29.5°	99.3°	151.1	264.1	129.8	02m25.2s
05:40	31°25.1'S	034°43.3'E	1.041	38.3°	92.7°	158.2	222.2	137.4	02m44.0s
05:50	34°14.9'S	040°42.3'E	1.043	45.3°	85.5°	162.7	201.2	143.0	02m59.5s
06:00	36°36.1'S	046°05.2'E	1.044	51.3°	77.2°	165.5	188.6	147.2	03m12.4s
06:10	38°35.1'S	051°09.6'E	1.045	56.5°	67.5°	167.4	180.5	150.4	03m23.1s
06:20	40°15.3'S	056°05.8'E	1.046	60.7°	55.9°	168.5	175.2	152.8	03m31.6s
06:30	41°38.2'S	060°59.9'E	1.046	64.1°	42.0°	169.2	171.8	154.4	03m37.9s
06:40	42°44.7'S	065°56.4'E	1.047	66.2°	25.7°	169.4	169.9	155.5	03m41.8s
06:50	43°34.7'S	070°58.3'E	1.047	67.0°	7.8°	169.3	169.3	155.9	03m43.5s
07:00	44°07.9'S	076°08.1'E	1.047	66.3°	349.7°	168.9	170.0	155.7	03m42.8s
07:10	44°23.7'S	081°28.0'E	1.047	64.3°	332.8°	168.1	171.9	154.8	03m39.7s
07:20	44°20.8'S	087°00.1'E	1.046	61.1°	318.0°	167.0	175.3	153.4	03m34.2s
07:30	43°57.7'S	092°47.1'E	1.045	56.9°	305.2°	165.4	180.7	151.2	03m26.3s
07:40	43°11.7'S	098°53.7'E	1.044	51.8°	294.0°	163.3	188.7	148.2	03m16.1s
07:50	41°59.0'S	105°27.5'E	1.043	45.9°	284.1°	160.4	201.1	144.3	03m03.4s
08:00	40°12.3'S	112°43.7'E	1.042	38.9°	274.9°	156.3	221.5	139.0	02m48.0s
08:10	37°36.4'S	121°17.8'E	1.039	30.3°	266.1°	150.1	261.2	131.7	02m29.3s
08:20	33°20.6'S	133°11.7'E	1.036	18.1°	256.6°	138.9	387.8	120.2	02m04.1s
08:25.6	26°20.4'S	151°09.5'E	1.030	0.0°	246.7°	118.3	-	101.7	01m32.1s

Table 2 lists the geographic coordinates of the central line for every ten minutes UTC. *Diameter Ratio* gives the Moon's apparent diameter divided by the Sun's. The Sun's altitude (*Alt*) and azimuth (*Azm*) are given along with the width of the path of totality (kilometers). The *Major* and *Minor* axes of the umbral shadow are also listed. Finally, the *Duration of Totality* is given (minutes and seconds).

5 – Local Circumstances for the Eclipse

Local circumstances for the ***Total Eclipse*** for 70 locations and cities are found in Table 3 (Africa) and Table 5 (Australia). All times are given in local time (see Table 1 for Time Zones, p. 9). The *Duration Totality* is the period between the beginning and end of totality. These times include the effects of the Moon's limb profile. The Sun's altitude is given at the instant of mid-totality.

Local circumstances for the ***Partial Eclipse*** for 71 cities are found in Table 4 (Africa) and Table 6 (Australia). All times are given in local time. The Sun's altitude is given at the instant of maximum eclipse. *Eclipse Magnitude* is given as the fraction of the Sun's *diameter* covered by the Moon at maximum eclipse.

The 2030 Total Solar Eclipse Circumstances Calculator is an interactive web page that can quickly calculate the local circumstances for the solar eclipse from any geographic location not included in Tables 3 through 6. The URL address for the calculator is:

https://www.EclipseWise.com/solar/SEcirc/2001-2100/SE2030Nov25Tcirc.html

Table 3 – Total Eclipse Circumstances for Africa

Location	Partial Eclipse Begins	Total Eclipse Begins	Total Eclipse Ends	Partial Eclipse Ends	Duration Of Totality	Sun Altitude
Botswana						
Kaa	06:25:42 am	07:21:42 am	07:23:47 am	08:26:20 am	02m06s	21°
Khisa	06:27:00 am	07:24:02 am	07:26:07 am	08:29:53 am	02m08s	24°
Mabuasehube	06:26:20 am	07:23:01 am	07:25:09 am	08:28:34 am	02m09s	23°
Makopong	06:26:33 am	07:23:41 am	07:25:43 am	08:29:38 am	02m02s	24°
Mabule	06:27:00 am	07:25:03 am	07:26:26 am	08:31:33 am	01m28s	25°
Omaweneno	06:27:04 am	07:24:06 am	07:26:06 am	08:29:50 am	02m01s	24°
Phepheng	06:26:43 am	07:23:43 am	07:25:54 am	08:29:42 am	02m12s	24°
Tsabong	06:27:23 am	07:24:48 am	07:25:52 am	08:30:01 am	01m04s	24°
Werda	06:26:28 am	07:23:56 am	07:25:30 am	08:29:49 am	01m37s	24°
Lesotho						
Butha-Buthe	06:30:50 am	07:30:47 am	07:32:32 am	08:39:50 am	01m44s	31°
Hlotse	06:30:56 am	07:31:22 am	07:32:04 am	08:39:43 am	00m50s	31°
Mokhotlong	06:31:35 am	07:32:07 am	07:33:36 am	08:41:31 am	01m26s	32°
Namibia						
Akanous	06:26:08 am	07:22:11 am	07:23:35 am	08:26:01 am	01m22s	21°
Arandis	06:24:52 am	07:18:59 am	07:20:07 am	08:20:04 am	01m05s	15°
Aranos	06:25:42 am	07:21:15 am	07:23:01 am	08:24:52 am	01m47s	20°
Karibib	06:24:09 am	07:18:14 am	07:19:57 am	08:19:59 am	01m43s	16°
Leonardville	06:25:05 am	07:20:16 am	07:22:18 am	08:23:49 am	02m02s	19°
Mile 108 on C34	06:24:15 am	07:17:30 am	07:19:19 am	08:18:17 am	01m52s	13°
Otjimbingwe	06:24:30 am	07:18:35 am	07:20:32 am	08:20:39 am	01m58s	16°
Rehoboth	06:25:14 am	07:20:04 am	07:21:30 am	08:22:25 am	01m26s	17°
Toscanini	06:23:49 am	07:16:56 am	07:18:35 am	08:17:21 am	01m38s	13°
Trekkopje	06:24:41 am	07:18:36 am	07:20:13 am	08:19:58 am	01m38s	15°
Uhlenhorst	06:25:26 am	07:20:34 am	07:22:09 am	08:23:30 am	01m35s	18°
Usakos	06:24:17 am	07:18:10 am	07:20:04 am	08:19:54 am	01m55s	15°
Windhoek	06:24:27 am	07:18:59 am	07:20:52 am	08:21:29 am	01m52s	17°
South Africa						
Ballito	06:32:17 am	07:33:19 am	07:35:50 am	08:44:35 am	02m31s	34°
Bethlehem	06:30:08 am	07:29:42 am	07:32:08 am	08:39:09 am	02m27s	30°
Bloemhof	06:29:14 am	07:28:29 am	07:29:07 am	08:35:19 am	00m47s	28°
Bothaville	06:28:57 am	07:27:42 am	07:30:03 am	08:36:04 am	02m23s	28°
Chatsworth	06:32:42 am	07:33:47 am	07:36:02 am	08:44:41 am	02m17s	34°
Durban	06:32:44 am	07:33:49 am	07:36:06 am	08:44:45 am	02m18s	34°
Edendale	06:32:15 am	07:33:03 am	07:35:14 am	08:43:34 am	02m12s	33°
eManzimtoti	06:32:54 am	07:34:04 am	07:36:04 am	08:44:52 am	01m59s	34°
Estcourt	06:31:11 am	07:31:18 am	07:33:21 am	08:40:54 am	02m02s	31°
Klerksdorp	06:28:27 am	07:27:27 am	07:29:19 am	08:35:31 am	01m52s	28°
Kroonstad	06:29:19 am	07:28:22 am	07:30:44 am	08:37:07 am	02m24s	29°
KwaDukuza	06:32:02 am	07:33:16 am	07:35:29 am	08:44:24 am	02m14s	34°
Mooi River	06:31:37 am	07:32:04 am	07:34:31 am	08:42:32 am	02m30s	33°
Morokweng	06:27:26 am	07:24:53 am	07:27:05 am	08:31:26 am	02m15s	25°
Phuthaditjhaba	06:30:35 am	07:30:25 am	07:32:51 am	08:40:11 am	02m29s	31°
Pieteraritzburg	06:32:12 am	07:32:58 am	07:35:11 am	08:43:29 am	02m14s	33°
Pinetown	06:32:38 am	07:33:39 am	07:35:53 am	08:44:28 am	02m15s	34°
Setlagole	06:27:36 am	07:25:47 am	07:27:45 am	08:32:59 am	01m58s	26°
uMnambithi	06:30:46 am	07:31:13 am	07:33:22 am	08:41:22 am	02m09s	32°
Vryburg	06:28:23 am	07:26:30 am	07:28:22 am	08:33:30 am	01m51s	26°
Welkom	06:29:42 am	07:28:49 am	07:30:41 am	08:37:02 am	01m52s	29°
Wolmaransstad	06:28:43 am	07:27:13 am	07:29:30 am	08:35:08 am	02m20s	28°

Table 4 - Partial Eclipse Circumstances for Africa

Country	City	Partial Eclipse Begins	Maximum Eclipse	Partial Eclipse Ends	Eclipse Magnitude	Sun Altitude
Angola	Luanda	-	06:07:25 am	07:02:23 am	0.688	6°
Botswana	Gaborone	06:25:55 am	07:25:11 am	08:31:39 am	0.961	26°
Burundi	Bujumbura	06:22:05 am	07:05:57 am	07:54:02 am	0.328	19°
Cen. Afr. Rep.	Bangui	-	06:00:32 am	06:38:12 am	0.239	5°
Chad	Ndjamena	-	-	06:16:18 am	0.047	0°
Congo	Brazzaville	-	06:04:26 am	06:55:49 am	0.534	6°
Dem. Rep. Congo	Kananga	06:17:29 am	07:05:44 am	07:58:55 am	0.493	13°
	Kinshasa	-	07:04:27 am	07:55:53 am	0.535	6°
	Kisangani	06:22:16 am	07:03:04 am	07:47:20 am	0.300	14°
	Kolwezi	06:17:33 am	07:10:06 am	08:08:41 am	0.590	19°
	Lubumbashi	06:18:09 am	07:11:38 am	08:11:26 am	0.591	21°
	Mbuji-Mayi	06:17:41 am	07:06:07 am	07:59:35 am	0.485	15°
Kenya	Mombasa	07:33:14 am	08:12:45 am	08:55:43 am	0.201	31°
	Nairobi	07:33:45 am	08:08:35 am	08:45:59 am	0.162	26°
Madagascar	Antananarivo	07:34:49 am	08:37:23 am	09:48:25 am	0.517	48°
	Fianarantsoa	07:35:00 am	08:40:02 am	09:53:58 am	0.594	49°
Malawi	Blantyre	06:22:17 am	07:20:01 am	08:25:16 am	0.603	32°
	Lilongwe	06:21:09 am	07:17:11 am	08:20:19 am	0.570	29°
Mozambique	Maputo	06:28:23 am	07:30:43 am	08:41:04 am	0.908	34°
Rwanda	Kigali	06:24:10 am	07:05:21 am	07:50:11 am	0.276	19°
South Africa	Alexandria	06:37:36 am	07:37:27 am	08:43:52 am	0.867	31°
	Benoni	06:27:53 am	07:28:30 am	08:36:36 am	0.969	30°
	Bloemfontein	06:31:13 am	07:31:05 am	08:37:57 am	0.970	29°
	Boksburg	06:27:43 am	07:28:13 am	08:36:11 am	0.969	29°
	Cape Town	06:38:48 am	07:34:28 am	08:35:39 am	0.771	24°
	Carletonville	06:27:50 am	07:27:59 am	08:35:29 am	0.984	29°
	Daveyton	06:27:41 am	07:28:16 am	08:36:19 am	0.965	30°
	East London	06:36:41 am	07:37:23 am	08:44:56 am	0.900	32°
	Evaton	06:28:02 am	07:28:27 am	08:36:16 am	0.981	29°
	Germiston	06:27:44 am	07:28:12 am	08:36:08 am	0.970	29°
	Johannesburg	06:27:44 am	07:28:08 am	08:35:59 am	0.973	29°
	Katlehong	06:27:50 am	07:28:20 am	08:36:16 am	0.973	29°
	Kimberley	06:30:34 am	07:29:46 am	08:35:50 am	0.965	27°
	Mamelodi	06:27:14 am	07:27:40 am	08:35:36 am	0.956	29°
	Mdantsana	06:36:34 am	07:37:09 am	08:44:35 am	0.899	32°
	Port Elizabeth	06:38:05 am	07:37:29 am	08:43:19 am	0.851	30°
	Pretoria	06:27:11 am	07:27:35 am	08:35:27 am	0.958	29°
	Maraisbu	06:27:40 am	07:28:01 am	08:35:47 am	0.972	29°
	Soweto	06:27:43 am	07:28:05 am	08:35:52 am	0.974	29°
	Tembisa	06:27:27 am	07:27:55 am	08:35:51 am	0.963	29°
	Vereeniging	06:28:11 am	07:28:38 am	08:36:30 am	0.983	29°
Tanzania	Dar-es-Salaam	07:28:52 am	08:14:38 am	09:05:10 am	0.287	32°
	Mwanza	07:26:14 am	08:07:01 am	08:51:25 am	0.254	22°
	Tabora	07:23:23 am	08:08:40 am	08:58:31 am	0.328	24°
	Zanzibar	07:29:26 am	08:14:01 am	09:03:05 am	0.270	31°
Uganda	Kampala	07:30:20 am	08:05:05 am	08:42:22 am	0.177	20°
Zambia	Kitwe	06:18:33 am	07:13:00 am	08:14:02 am	0.613	23°
	Lusaka	06:19:18 am	07:15:35 am	08:18:50 am	0.684	24°
	Ndola	06:18:42 am	07:13:20 am	08:14:34 am	0.611	23°
Zimbabwe	Bulawayo	06:22:04 am	07:20:55 am	08:27:13 am	0.807	27°
	Harare	06:21:15 am	07:19:38 am	08:25:31 am	0.713	28°

Table 5 – Total Eclipse Circumstances for Australia

Location	Partial Eclipse Begins	Total Eclipse Begins	Total Eclipse Ends	Partial Eclipse Ends	Duration of Totality	Sun Altitude
South Australia						
Baird Bay	05:48:56 pm	06:49:31 pm	06:51:22 pm	07:46:14 pm	01m53s	17°
Calca	05:49:05 pm	06:49:33 pm	06:51:34 pm	07:46:21 pm	02m00s	17°
Hawker	05:52:40 pm	06:51:43 pm	06:52:44 pm	07:46:29 pm	01m00s	13°
Minnipa	05:49:47 pm	06:49:59 pm	06:51:48 pm	07:46:23 pm	01m52s	16°
Parachilna	05:53:29 pm	06:52:07 pm	06:53:51 pm	07:47:13 pm	01m47s	13°
Poochera	05:49:43 pm	06:49:57 pm	06:51:57 pm	07:46:34 pm	02m01s	16°
Port Kenny	05:49:08 pm	06:49:41 pm	06:51:19 pm	07:46:09 pm	01m39s	17°
Nilpena	05:53:31 pm	06:52:20 pm	06:53:52 pm	07:47:21 pm	01m32s	13°
Streaky Bay	05:49:14 pm	06:49:48 pm	06:51:43 pm	07:46:37 pm	01m59s	17°
Wilpena Pound	05:53:08 pm	06:51:46 pm	06:53:28 pm	07:46:46 pm	01m44s	13°
Wudinna	05:49:46 pm	06:50:08 pm	06:51:18 pm	07:46:08 pm	01m10s	16°
Yaninee	05:49:44 pm	06:50:02 pm	06:51:37 pm	07:46:15 pm	01m37s	16°
New South Wales						
Barringun	06:28:17 pm	07:24:11 pm	07:25:02 pm	-	00m50s	6°
Milparinka	06:26:25 pm	07:23:31 pm	07:25:08 pm	-	01m39s	9°
Tibooburra	06:26:47 pm	07:24:02 pm	07:25:12 pm	-	01m08s	9°
Wanaaring	06:27:10 pm	07:23:55 pm	07:24:19 pm	-	00m32s	7°
Queensland						
Bollon	05:29:34 pm	06:24:25 pm	06:25:52 pm	-	01m29s	4°
Chinchilla	05:31:07 pm	06:24:42 pm	06:26:08 pm	-	01m28s	1°
Condamine	05:30:55 pm	06:24:41 pm	06:26:08 pm	-	01m29s	1°
Cunnamulla	05:29:11 pm	06:24:47 pm	06:26:03 pm	-	01m15s	5°
Eulo	05:28:57 pm	06:25:07 pm	06:25:38 pm	-	00m39s	6°
Hungerford	05:27:58 pm	06:23:55 pm	06:25:38 pm	-	01m43s	7°
Miles	05:31:09 pm	06:24:50 pm	06:26:23 pm	-	01m33s	1°
Surat	05:30:36 pm	06:24:42 pm	06:26:16 pm	-	01m35s	2°
Yuleba	05:31:08 pm	06:25:19 pm	06:26:25 pm	-	01m07s	2°

Table 6 - Partial Eclipse Circumstances for Australia

Location	Partial Eclipse Begins	Maximum Eclipse	Partial Eclipse Ends	Eclipse Magnitude	Sun Altitude
Adelaide, SA	05:49:36 pm	06:49:02 pm	07:43:20 pm	0.925	14°
Albury, VIC	06:22:10 pm	07:17:31 pm	-	0.816	9°
Ballarat, VIC	06:19:40 pm	07:16:09 pm	08:08:16 pm	0.810	12°
Bendigo, VIC	06:20:34 pm	07:17:01 pm	08:09:04 pm	0.825	11°
Brisbane, QLD	05:30:36 pm	06:24:04 pm	-	0.965	0°
Cairns, QLD	05:41:05 pm	06:33:33 pm	-	0.703	0°
Canberra, ACT	06:23:31 pm	07:18:00 pm	-	0.814	7°
Darwin, NT	05:17:50 pm	06:06:59 pm	06:51:50 pm	0.406	11°
Geelong, VIC	06:19:23 pm	07:15:30 pm	08:07:21 pm	0.792	12°
Hobart, TAS	06:16:40 pm	07:09:48 pm	07:59:23 pm	0.659	12°
Launceston, TAS	06:17:45 pm	07:11:32 pm	08:01:36 pm	0.692	11°
Melbourne, VIC	06:19:55 pm	07:15:49 pm	08:07:28 pm	0.794	11°
Newcastle, NSW	06:26:00 pm	07:19:43 pm	-	0.845	3°
Perth, WA	03:02:04 pm	04:12:33 pm	05:15:13 pm	0.813	34°
Southport, QLD	05:30:12 pm	06:23:32 pm	-	0.947	0°
Sydney, NSW	06:25:09 pm	07:18:58 pm	-	0.828	4°
Toowoomba, QLD	05:30:29 pm	06:24:22 pm	-	0.976	1°
Townsville, QLD	05:38:22 pm	06:31:53 pm	-	0.780	0°
Wollongong, NSW	06:24:38 pm	07:18:30 pm	-	0.818	5°

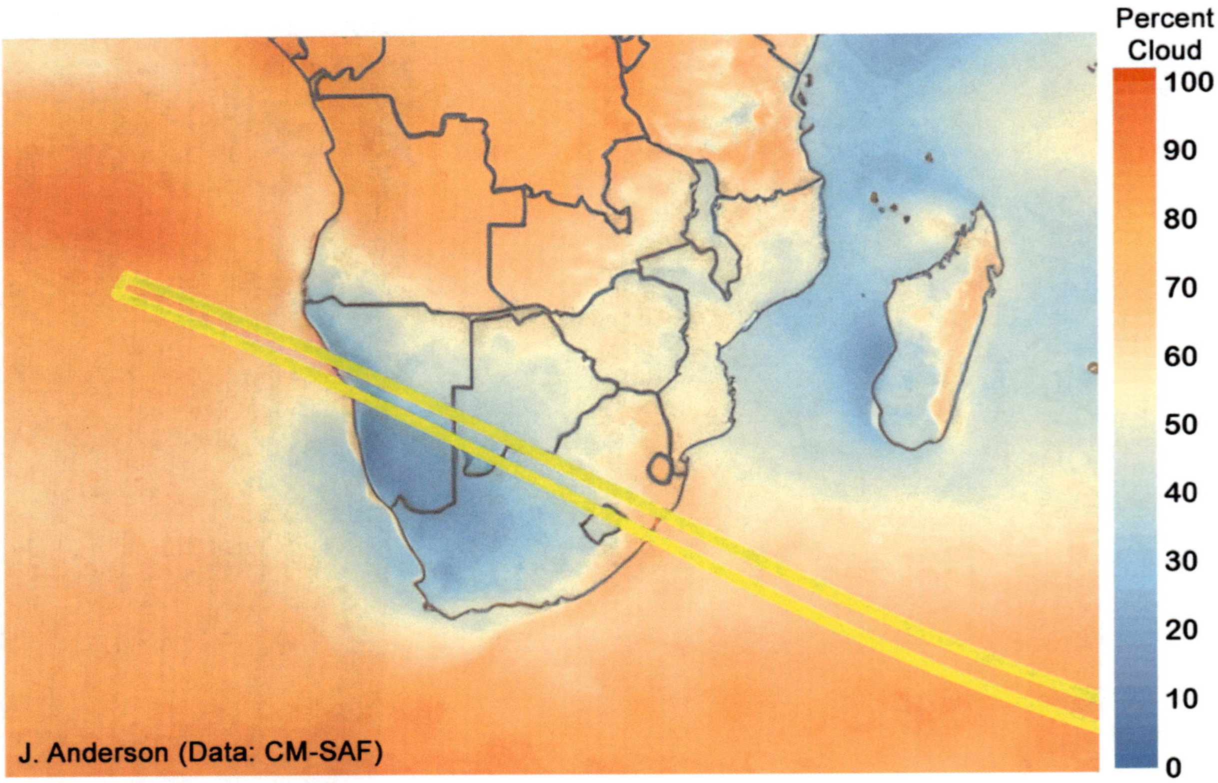

Figure 9 – 2030 total solar eclipse through Africa with average November daytime cloud cover derived from 20 years of satellite data, 2000-2019. Data: CM-SAF. (map by Jay Anderson)

6 – Weather Prospects for the 2030 Total Solar Eclipse Through Africa

The 2030 total eclipse track crosses southern Africa including Namibia, Botswana, South Africa, and Lesotho. The western half of this track offers excellent prospects for clear sky during the eclipse with the probability of cloud increases to nearly 70% in Durban. Table 7 lists eleven locations in the path of totality and include the duration of totality, the average high temperature, and the average cloud cover during November.

Table 7 – Africa: Average High Temperature and Cloud Cover in November

Country	City	Duration Totality	Ave. High Temp.[1]	Ave. Cloud Cover[2]
Namibia	Mile 108 (C34)	1m 52s	19°C (67°F)	18%
Namibia	Windhoek	1m 53s	30°C (86°F)	23%
Namibia	Leonardville	2m 03s	32°C (89°F)	19%
Botswana	Tsabong	1m 05s	32°C (90°F)	24%
South Africa	Klerksdorp	1m 31s	29°C (84°F)	34%
South Africa	Kroonstad	2m 24s	27°C (81°F)	36%
South Africa	Bethlehem	2m 28s	24°C (76°F)	43%
South Africa	Phuthaditjhaba	2m 29s	23°C (73°F)	40%
South Africa	Estcourt	2m 32s	26°C (78°F)	54%
South Africa	Pietermaritzburg	2m 22s	27°C (80°F)	62%
South Africa	Durban	2m 26s	26°C (78°F)	67%

[1]*Average High Temperature in November [source: www.weatherspark.com]*

[2]*Average Cloud Cover in November (since 2000) [source: www.timeanddate.com]*

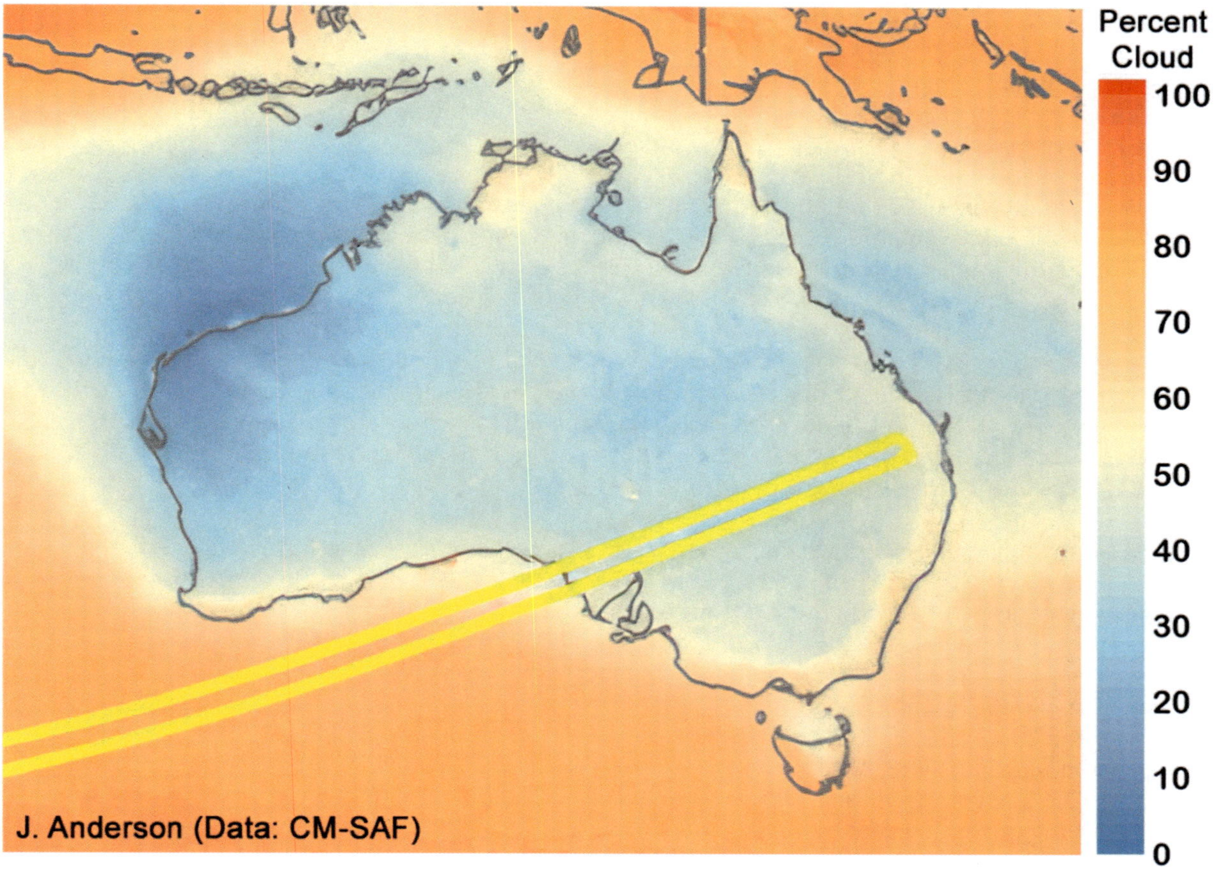

Figure 10 – 2030 total solar eclipse through Australia with average November daytime cloud cover derived from 20 years of satellite data, 2000-2019. Data: CM-SAF. (map by Jay Anderson)

7 – Weather Prospects for the 2030 Total Solar Eclipse Through Australia

Although the 2030 eclipse occurs at the traditional start of Australia's wet season, the track passes through a region offering good chances for clear sky during the eclipse (probability of cloud amount ranges from 29% to 49%). Table 8 lists ten locations in the path of totality and include the duration of totality, the average high temperature, and the average cloud cover during November.

Table 8 – Australia: Average High Temperature and Cloud Cover in November

Country	City	Duration Totality	Ave. High Temp.[1]	Ave. Cloud Cover[2]
Australia	Scale Bay	2m 02s	31°C (87°F)	49%
Australia	Maryvale	2m 02s	28°C (82°F)	43%
Australia	Poochera	2m 01s	33°C (91°F)	38%
Australia	Oakden Hills	1m 56s	xx°C (xx°F)	37%
Australia	Parachilna	1m 47s	33°C (91°F)	42%
Australia	Tibooburra	1m 05s	31°C (88°F)	29%
Australia	Currawinya NP	1m 44s	29°C (85°F)	37%
Australia	Cunnamulla	1m 14s	31°C (88°F)	34%
Australia	Surat	1m 35s	30°C (86°F)	34%
Australia	Miles	1m 32s	31°C (87°F)	39%

[1]Average High Temperature in November [source: www.weatherspark.com]
[2]Average Cloud Cover in November (since 2000) [source: www.timeanddate.com]

8 – Detailed Maps of the Path of Totality

A series of 17 full-page maps (pages 23-39) cover the entire 2030 path of totality through Africa and Australia. The *Table of Contents* (page iii) can be used to quickly navigate to the map of interest since it lists the region featured on each map. North is up in each map.

The map scale is approximately 1:1,400,000, which corresponds to 1 inch ≈ 22 miles (1 cm ≈ 14 km).[5] This large scale shows both major and minor roads, towns and cities, rivers, lakes, parks, national forests, wilderness areas and mountain ranges. Two reference scales for 50 miles and 50 kilometers appear at the bottom of each map.

The path of totality on each map is depicted as a lightly shaded region with the northern and southern limits clearly labeled. The total eclipse can be seen only inside this path (a partial eclipse is visible outside the path). The closer one gets to the central line, the longer the total eclipse lasts. Gray lines inside the path mark the duration of the total eclipse in 30-second steps. This makes it easy to estimate the duration of totality from any location in the eclipse path.

The local time of mid-eclipse is marked by a series of white lines crossing the eclipse path every five minutes. Abbreviations for local times are CAT, SAST, ACDT, AEDT, and AEST (see Table 1, p. 9). Eclipse circumstances on the central line are labeled with the local time of mid-eclipse, duration of totality (minutes and seconds) and altitude of the Sun.

All maps were produced using Google Maps as the underlying map with overlying eclipse graphics generated using JavaScript code. A web page is available to the user for examining any part of the 2030 eclipse path at a range of zoom magnifications. An added benefit of the web page is that it automatically calculates the local circumstances for any point chosen by the user, with just the click of a mouse. For more information and to access the interactive 2030 eclipse path plotted on Google Maps, visit:

https://www.eclipsewise.com/solar/SEgmapx/2001-2100/SE2030Nov25Tgmapx.html#map

9 – EclipseWise.com Web Site

EclipseWise.com has individual web pages, maps and diagrams for every solar and lunar eclipse covering the 5000-year period from 2000 BCE to 3000 CE. This includes 11,898 solar eclipses and 12,064 lunar eclipses. Much of the design, layout and graphics were inspired by the recent publications *Thousand Year Canon of Solar Eclipses 1501 to 2500* and the *Thousand Year Canon of Lunar Eclipses 1501 to 2500.* For more information on these books, see: ***https://www.astropixels.com/pubs***

The graphical user interface used by *EclipseWise.com* offers an intuitive way of accessing eclipse predictions. For example, the home page presents a concise preview of all upcoming solar and lunar eclipses over several years. Each small eclipse diagram gives a quick preview of an eclipse and links to a dedicated page for that eclipse.

The main or top pages of EclipseWise.com are:

Home Page (both solar and lunar eclipses): ***https://www. eclipsewise.com/eclipse.html***
Solar Eclipses Page: ***https://www. eclipsewise.com/solar/solar.html***
Lunar Eclipses Page: ***https://www. eclipsewise.com/lunar/lunar.html***

---

[5] Because of the Mercator map projection, the actual scale on a given map can vary by up to 10% of this value.

10 – EclipseWise.com and the 2030 Eclipse

EclipseWise.com has a series of pages and resources devoted to the 2030 eclipse. The main page is located at:

https://www.eclipsewise.com/solar/SEprime/2001-2100/SE2030Nov25Tprime.html

It provides links to detailed eclipse path maps, tables of eclipse path coordinates, Besselian elements, and more. The link to an interactive Google Map with the eclipse path plotted on it allows the user to zoom into a part of the path. Click on any point on the map to display the eclipse circumstances and duration of totality at that location.

Other features include information on eye safety, eclipse photography, the sky during totality, and additional data tables about the eclipse path. This web site will continue to add features as the eclipse approaches.

The *2030 Total Solar Eclipse Circumstances Calculator* is an interactive web page that can quickly calculate the local circumstances for the eclipse from any geographic location. The *Calculator* is located at:

https://www.eclipsewise.com/solar/SEcirc/2001-2100/SE2030Nov25Tcirc.html

11 – Eclipse Predictions

The algorithms and software for the eclipse predictions were developed primarily from the *Explanatory Supplement to the Astronomical Ephemeris* (Her Majesty's Nautical Almanac Office, 1974) with additional algorithms from *Elements of Solar Eclipses: 1951–2200* (Meeus, 1989). The solar and lunar ephemerides were generated from the JPL DE430. All eclipse calculations were made using a value for the Moon's radius of k=0.2722810 for the path of totality. Eclipse contact times for locations and cities (Tables 3 and 5) have been corrected for the effects of the Moon's limb profile. A value for ΔT of 69.2 seconds was used to convert the predictions from Terrestrial Time (TT) to Coordinated Universal Time (UTC).

Photo 4 – Viewing the Total Solar Eclipse of August 21, 2017.
Casper, Wyoming, © 2017 F. Espenak, www.MrEclipse.com

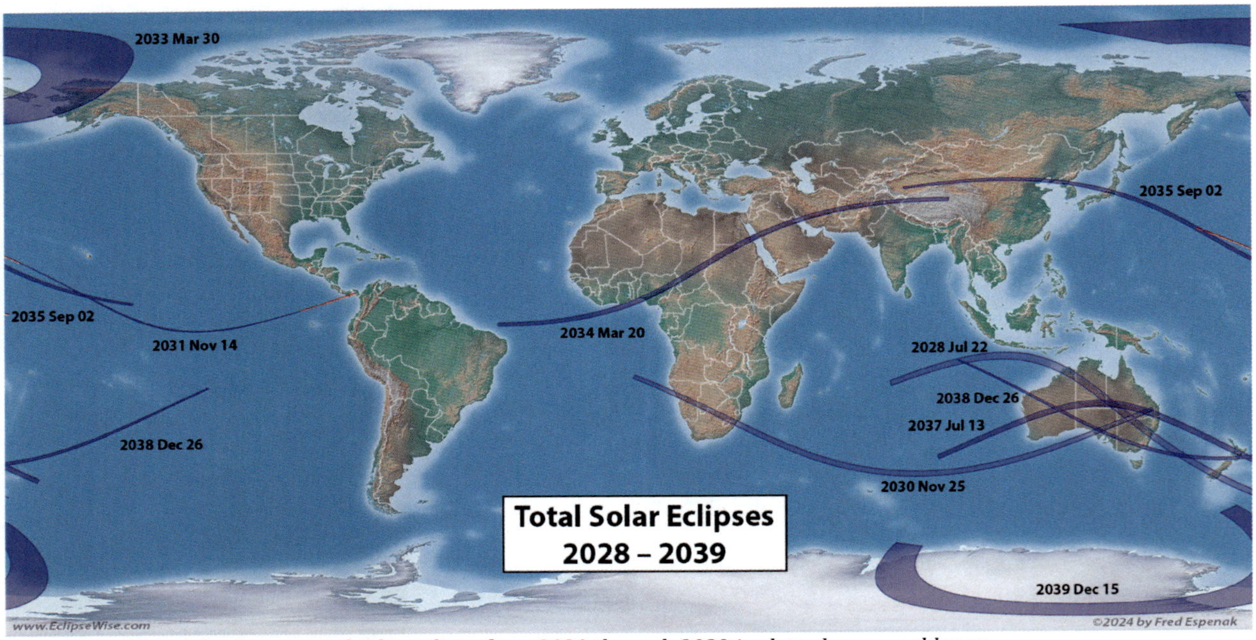

Figure 11 – The path of every total solar eclipse from 2030 through 2039 is plotted on a world map.

12 – Total Solar Eclipses: 2028 – 2039

During the 12-year period 2028 through 2039, Earth will experience nine total solar eclipses. Examination of the map in Figure 11 shows that none of them are visible from North America (except 2033 that crosses Alaska). In contrast, Australia is fortunate to have no less than four total solar eclipses during the same period (2028 July 22, 2030, Nov 25, 2037 July 13, and 2038 Dec 26). Information for each eclipse appears in *Table 9*.

Table 9 – Data for Total Solar Eclipses

Calendar Date	Time of Greatest Eclipse	Saros Number	Latitude	Longitude	Sun Alt	Path Width km	Duration Totality
2028 Jul 22	02:55:29	146	16°S	127°E	53°	230	05m10s
2030 Nov 25	06:50:27	133	44°S	71°E	67°	169	03m44s
2031 Nov 14*	21:06:20	143	1°S	138°W	72°	38	01m08s
2033 Mar 30	18:01:25	120	71°N	156°W	11°	781	02m37s
2034 Mar 20	10:17:35	130	16°N	22°E	73°	159	04m09s
2035 Sep 02	01:55:36	145	29°N	158°E	68°	116	02m54s
2037 Jul 13	02:39:25	127	25°S	139°E	43°	201	03m58s
2038 Dec 26	00:58:58	142	40°S	164°E	73°	95	02m18s
2039 Dec 15	16:22:34	152	81°S	173°E	18°	380	01m51s

* The eclipse of 2031 Nov 14 is a Hybrid Eclipse or Annular/Total Eclipse.

Explanation of Column Headings in Table 9

Calendar Date – The UTC date of the eclipse at the instant of Greatest Eclipse
Time of Greatest Eclipse – The UTC Time of Greatest Eclipse
Saros Number – The number of the Saros series that the eclipse belongs to
Latitude – The latitude of the Moon's umbral shadow at instant of Greatest Eclipse
Longitude – The longitude of the Moon's umbral shadow at instant of Greatest Eclipse
Sun Alt – The altitude of the Sun at the instant of Greatest Eclipse
Path Width – The width of the path of totality (km) at the instant of Greatest Eclipse
Duration Totality – The central line duration of totality at the instant of Greatest Eclipse

Total Solar Eclipses from 2028 to 2039

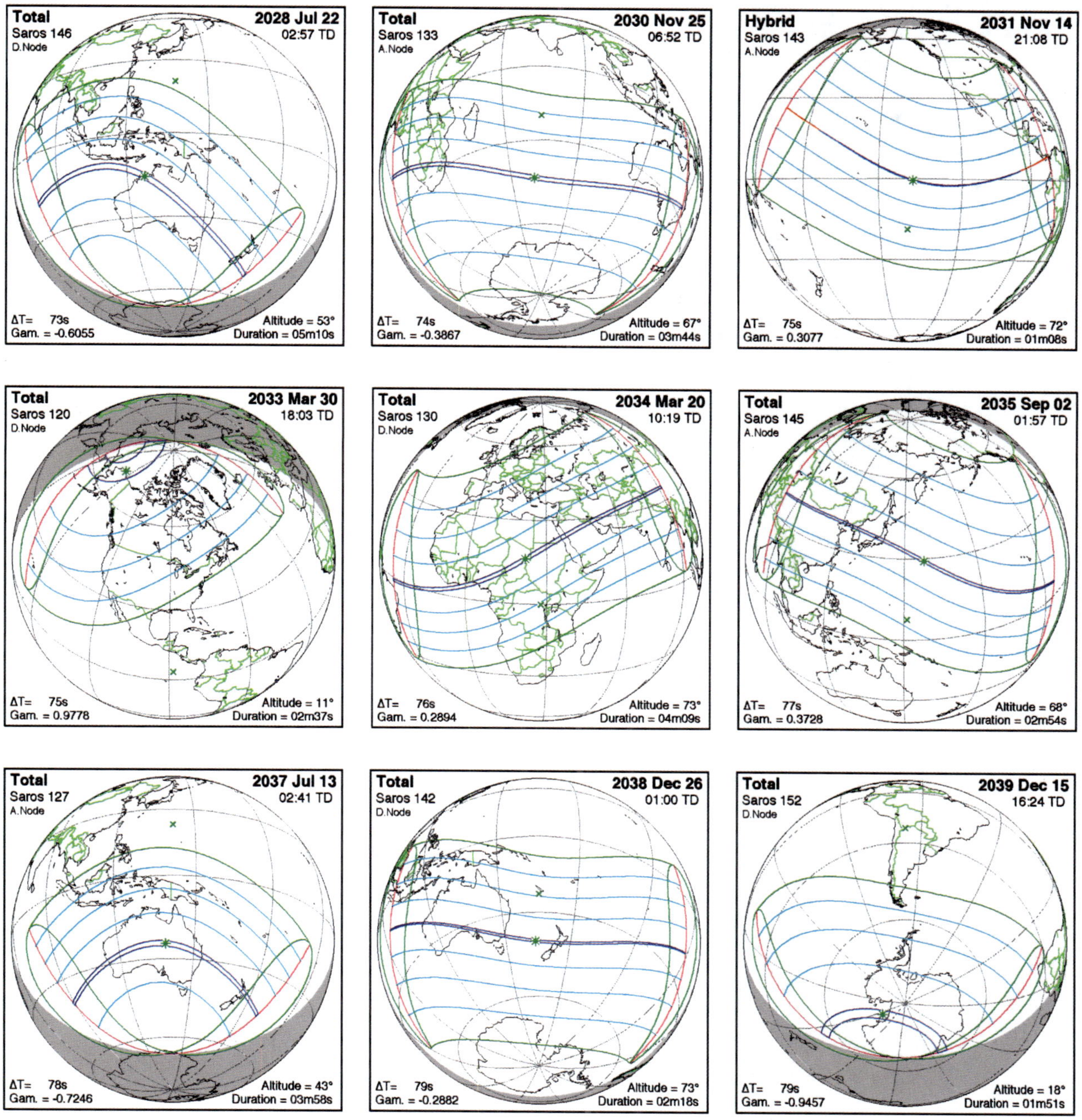

Figure 12 – Total Solar Eclipses from 2028 through 2039.
Courtesy of ***Thousand Year Canon of Solar Eclipses1501 to 2500****, F. Espenak, (2014)*

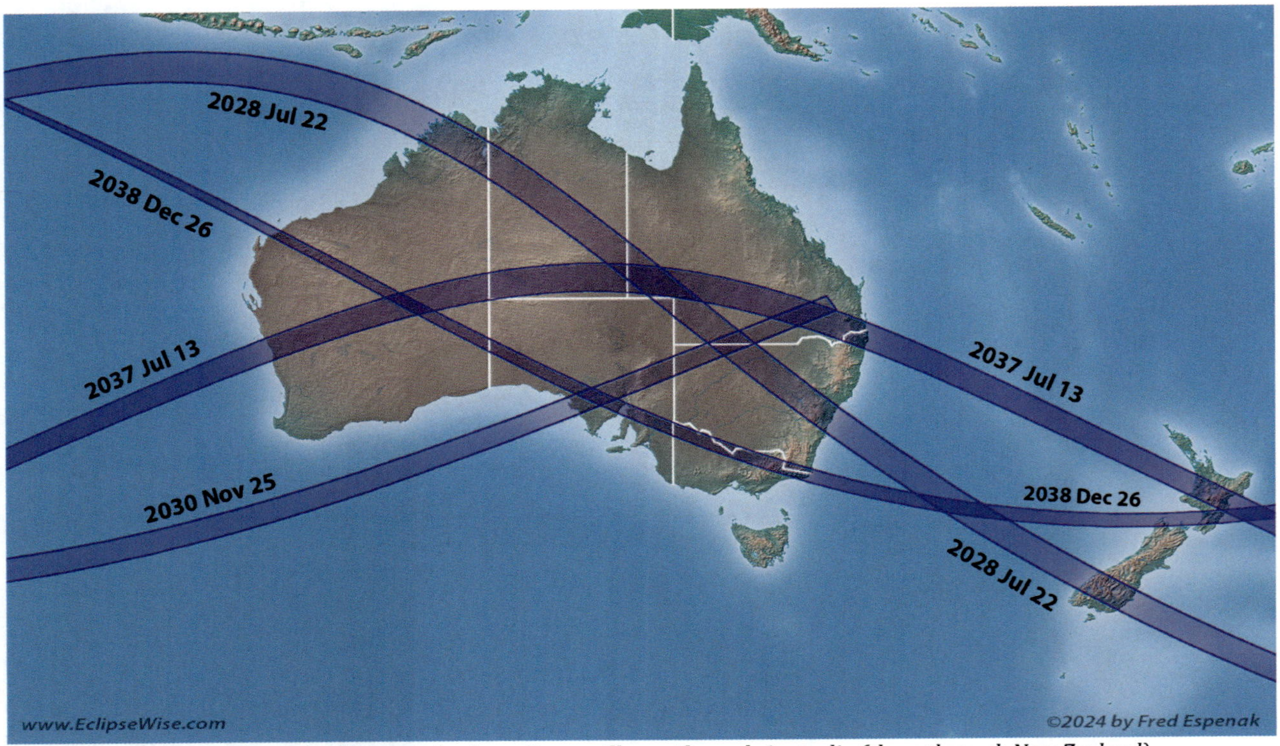

Figure 13 – The path of four total solar eclipses will pass through Australia (three through New Zealand) between 2028 and 2038.

13 – Total Solar Eclipses in Australia and New Zealand: 2028 – 2039

During the 11-year period 2028 through 2038, the path of four total solar eclipses will pass through Australia. The path of three of these eclipses also cross New Zealand.

1) July 22, 2028† - Path crosses the Kimberley in Western Australia, Northern Territory, southwest Queensland, New South Wales and passing centrally through Sydney. Maximum Duration in Australia: 5m 10s

2) November 25, 2030 - Path crosses Africa, South Australia, northwest New South Wales, southern Queensland ending at sunset in southeast Queensland. Maximum Duration in Australia: 2m 02s

3) July 13, 2037† - Path crosses southern Western Australia, southern Northern Territory, western Queensland and finally passing over Brisbane and the Gold Coast. Maximum Duration in Australia: 3m 59s

4) December 26, 2038† - Path crosses central Western Australia, South Australia, and along the New South Wales/Victoria border. Maximum Duration in Australia: 2m 10s

---

† Path of totality also crosses New Zealand

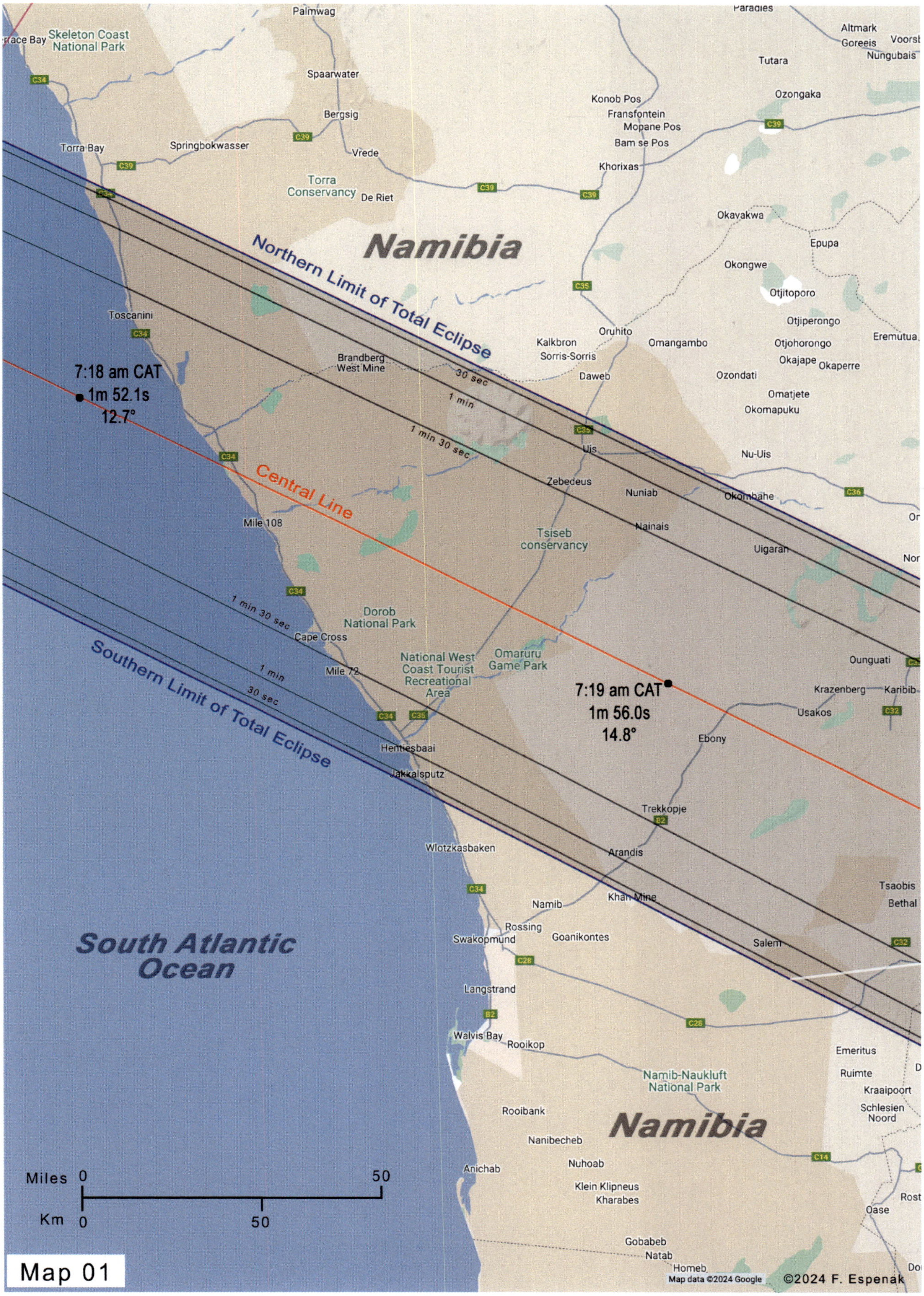
Namibia
Northern Limit of Total Eclipse
Southern Limit of Total Eclipse
Central Line
30 sec
1 min
1 min 30 sec
7:18 am CAT
1m 52.1s
12.7°
7:19 am CAT
1m 56.0s
14.8°
South Atlantic Ocean
Skeleton Coast National Park
Torra Conservancy
Dorob National Park
National West Coast Tourist Recreational Area
Omaruru Game Park
Tsiseb conservancy
Namib-Naukluft National Park
Palmwag
Spaarwater
Bergsig
Springbokwasser
Torra Bay
Vrede
De Riet
Toscanini
Brandberg West Mine
Mile 108
Cape Cross
Mile 72
Hentiesbaai
Jakkalsputz
Wlotzkasbaken
Swakopmund
Langstrand
Walvis Bay
Rooikop
Rooibank
Nanibecheb
Anichab
Nuhoab
Klein Klipneus
Kharabes
Gobabeb
Natab
Homeb
Namib
Rossing
Goanikontes
Khan Mine
Arandis
Trekkopje
Ebony
Salem
Usakos
Karibib
Krazenberg
Ounguati
Tsaobis
Bethal
Emeritus
Ruimte
Kraaipoort
Schlesien Noord
Oase
Zebedeus
Uis
Nuniab
Nainais
Uigaran
Okombahe
Nu-Uis
Daweb
Kalkbron
Sorris-Sorris
Oruhito
Omangambo
Ozondati
Omatjete
Okomapuku
Okajape
Okaperre
Otjohorongo
Otjiperongo
Otjitoporo
Okongwe
Epupa
Okavakwa
Eremutua
Khorixas
Bam se Pos
Mopane Pos
Fransfontein
Konob Pos
Ozongaka
Tutara
Altmark
Goreeis
Nungubais
Miles 0 50
Km 0 50
Map 01
Map data ©2024 Google
©2024 F. Espenak

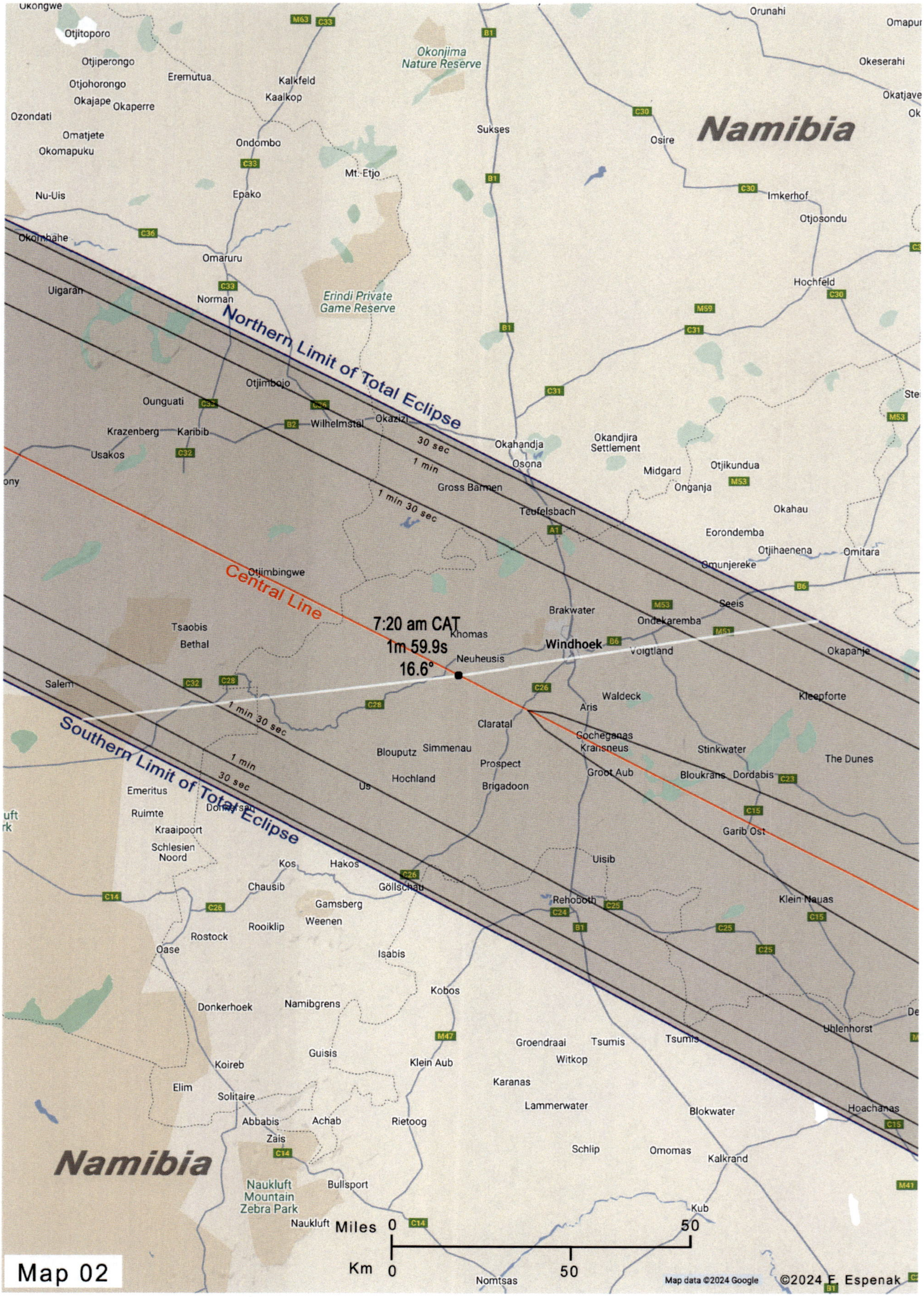
Namibia
Northern Limit of Total Eclipse
Central Line
Southern Limit of Total Eclipse
30 sec
1 min
1 min 30 sec
1 min 30 sec
1 min
30 sec
7:20 am CAT
1m 59.9s
16.6°
Windhoek
Okonjima Nature Reserve
Erindi Private Game Reserve
Naukluft Mountain Zebra Park
Otjitoporo
Otjiperongo
Otjohorongo
Okajape
Okaperre
Ozondati
Omatjete
Okomapuku
Eremutua
Kalkfeld
Kaalkop
Ondombo
Mt. Etjo
Sukses
Osire
Orunahi
Okeserahi
Nu-Uis
Epako
Imkerhof
Otjosondu
Omaruru
Uigarab
Norman
Hochfeld
Otjimbojo
Ounguati
Krazenberg
Karibib
Usakos
Wilhelmstal
Okazizi
Okahandja
Osona
Okandjira Settlement
Midgard
Onganja
Otjikundua
Gross Barmen
Teufelsbach
Okahau
Eorondemba
Otjihaenena
Omitara
Omunjereke
Otjimbingwe
Seeis
Brakwater
Ondekaremba
Tsaobis
Bethal
Khomas
Neuheusis
Voigtland
Okapanje
Salem
Aris
Waldeck
Kleepforte
Claratal
Gocheganas
Kransneus
Stinkwater
Blouputz
Simmenau
Prospect
The Dunes
Hochland
Brigadoon
Groot Aub
Bloukrans
Dordabis
Emeritus
Us
Ruimte
Kraaipoort
Schlesien Noord
Garib Ost
Kos
Hakos
Uisib
Chausib
Göllschau
Rehoboth
Klein Nauas
Gamsberg
Weenen
Rooiklip
Rostock
Oase
Isabis
Kobos
Donkerhoek
Namibgrens
Uhlenhorst
Groendraai
Tsumis
Witkop
Guisis
Koireb
Klein Aub
Karanas
Elim
Solitaire
Lammerwater
Abbabis
Achab
Rietoog
Blokwater
Hoachanas
Zais
Schlip
Omomas
Kalkrand
Bullsport
Naukluft
Kub
Nomtsas
Miles 0 50
Km 0 50
Map 02
Map data ©2024 Google
©2024 F. Espenak

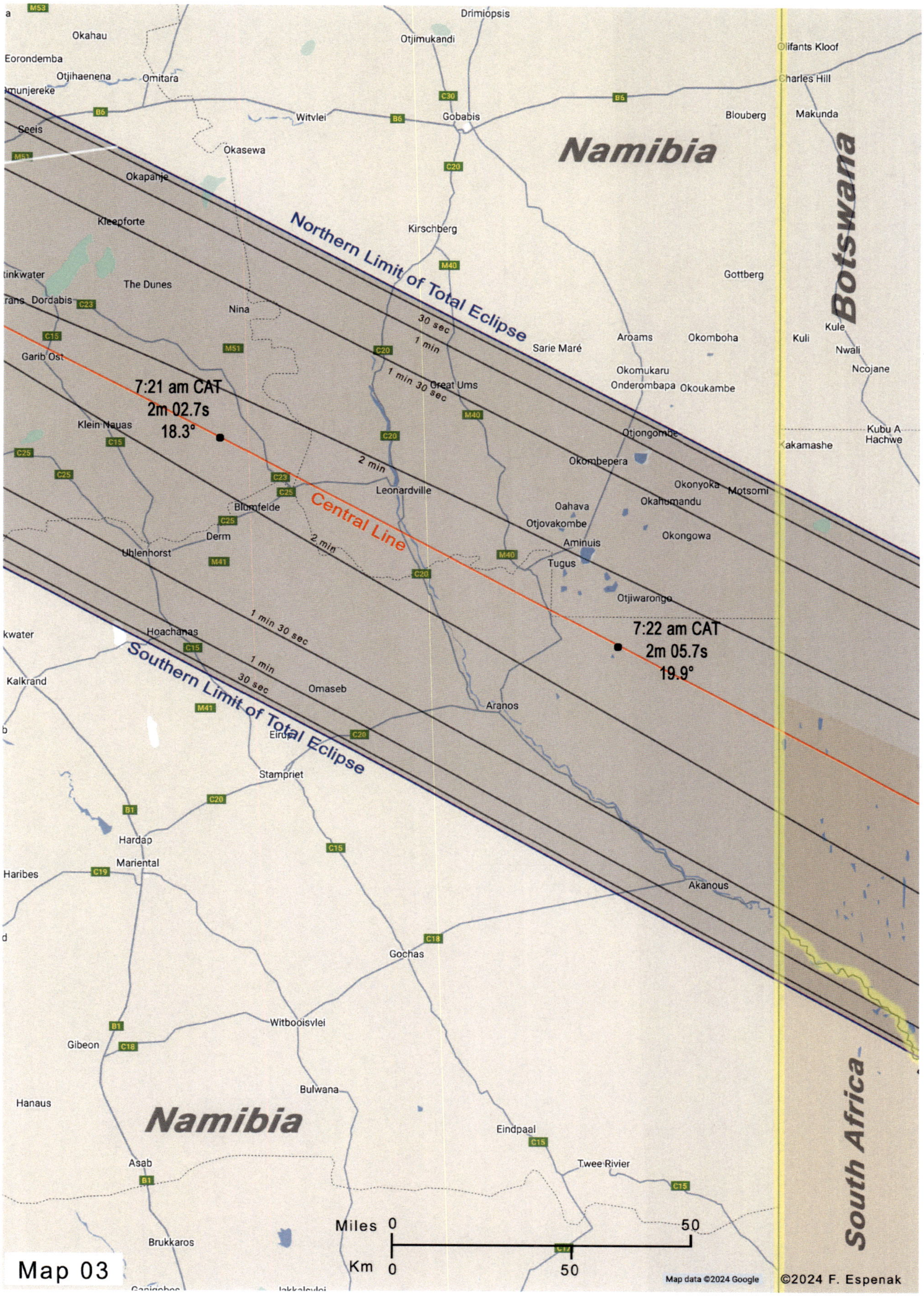
Namibia
Botswana
South Africa
Northern Limit of Total Eclipse
Southern Limit of Total Eclipse
Central Line
7:21 am CAT
2m 02.7s
18.3°
7:22 am CAT
2m 05.7s
19.9°
30 sec
1 min
1 min 30 sec
2 min
Gobabis
Witvlei
Leonardville
Aranos
Stampriet
Gochas
Mariental
Hardap
Omaseb
Otjiwarongo
Akanous
Miles 0 50
Km 0 50
Map data ©2024 Google
©2024 F. Espenak
Map 03

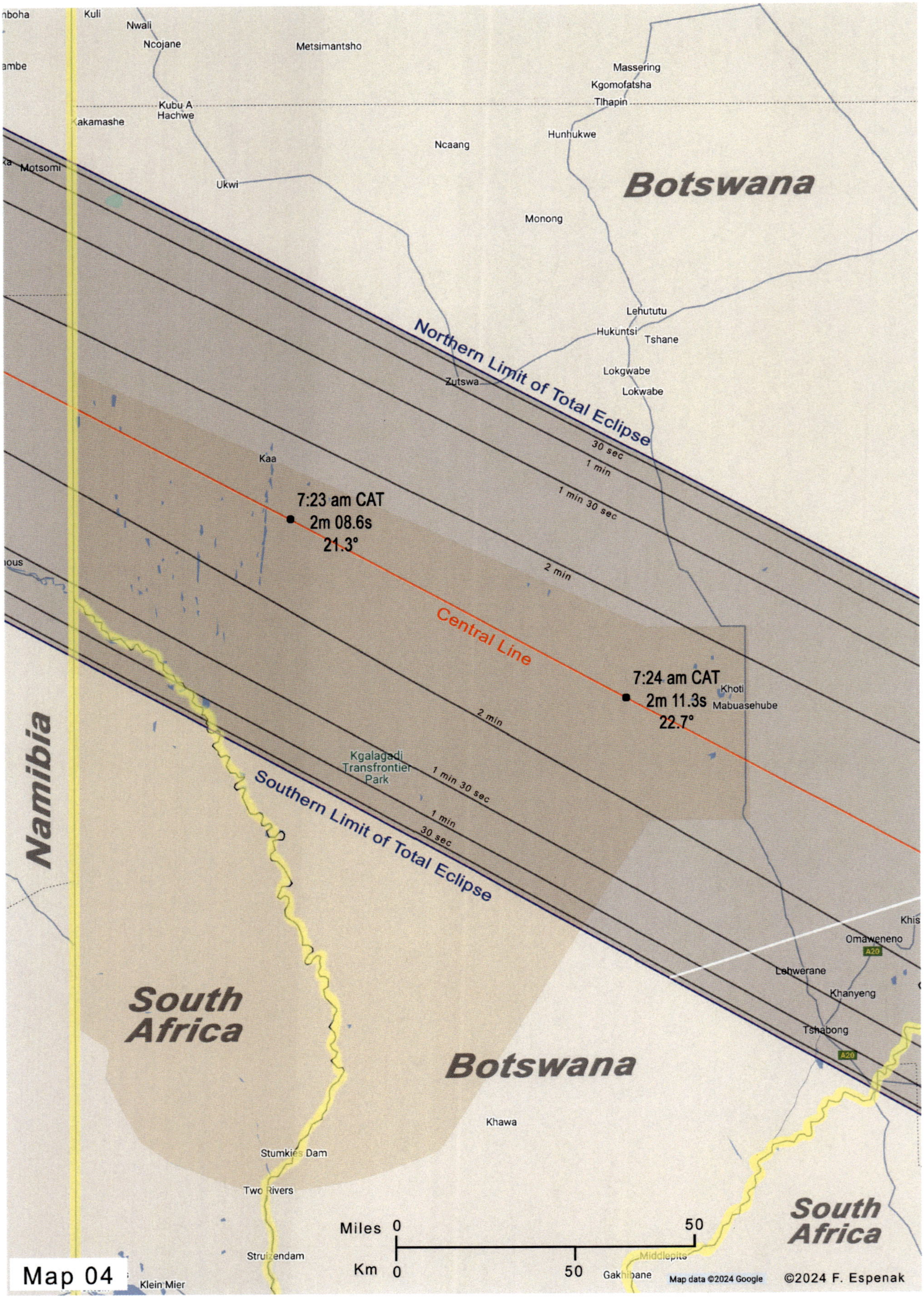
Kuli
Nwali
Ncojane
Metsimantsho
Massering
Kgomofatsha
Tlhapin
Kubu A
Hachwe
Ncaang
Hunhukwe
Botswana
Motsomi
Ukwi
Monong
Lehututu
Hukuntsi
Tshane
Lokgwabe
Lokwabe
Zutswa
Northern Limit of Total Eclipse
30 sec
1 min
1 min 30 sec
Kaa
7:23 am CAT
2m 08.6s
21.3°
2 min
Central Line
7:24 am CAT
2m 11.3s
22.7°
Khoti
Mabuasehube
Namibia
2 min
Kgalagadi
Transfrontier
Park
1 min 30 sec
1 min
30 sec
Southern Limit of Total Eclipse
Khis
Omaweneno
Lehwerane
Khanyeng
Tshabong
South
Africa
Botswana
Khawa
Stumkies Dam
Two Rivers
South
Africa
Miles 0
50
Km 0
50
Struizendam
Middlepits
Map 04
Klein Mier
Gakhibane
Map data ©2024 Google
©2024 F. Espenak

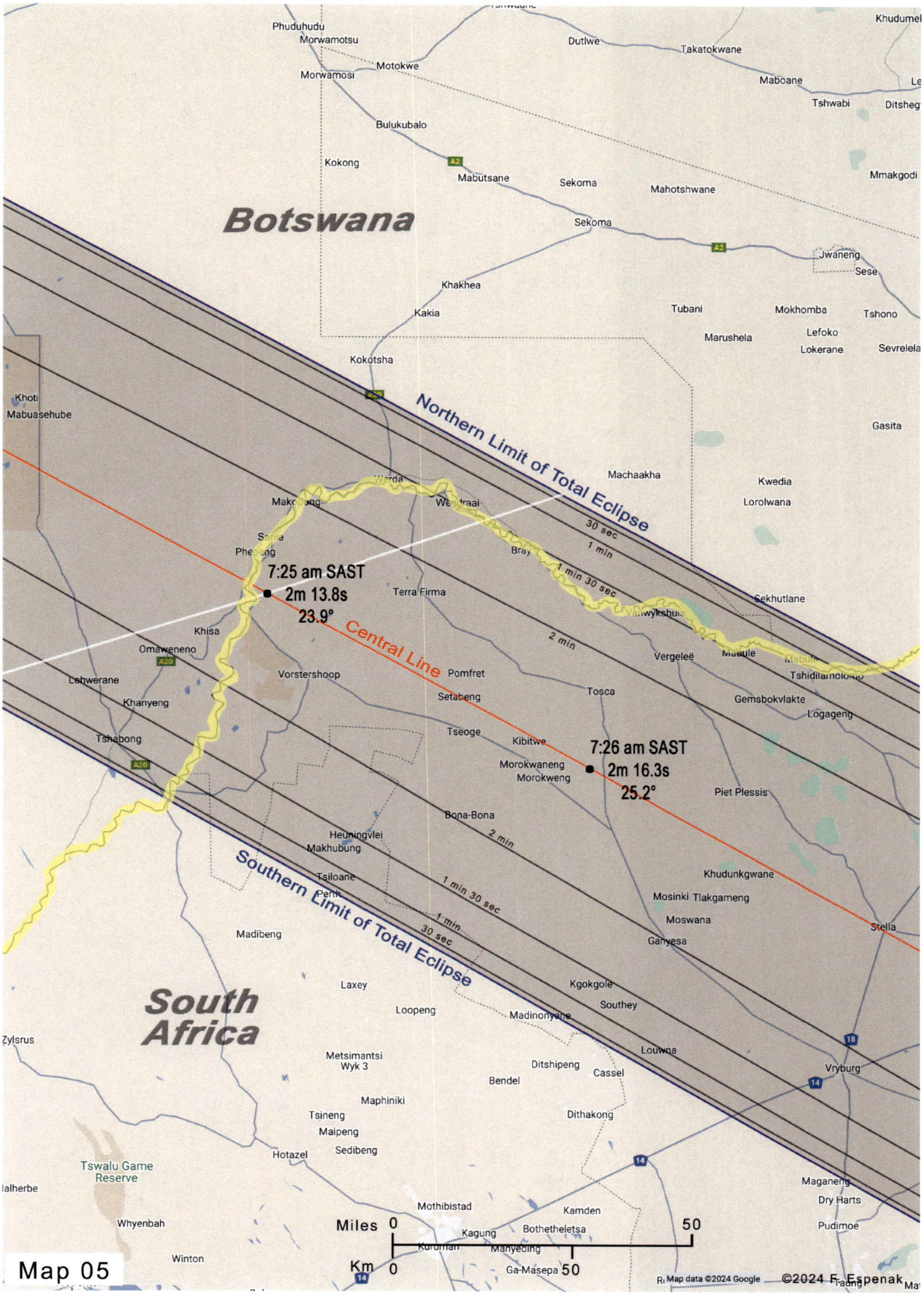
Botswana
South Africa
Northern Limit of Total Eclipse
Southern Limit of Total Eclipse
Central Line
30 sec
1 min
1 min 30 sec
2 min
2 min
1 min 30 sec
1 min
30 sec
7:25 am SAST
2m 13.8s
23.9°
7:26 am SAST
2m 16.3s
25.2°
Phuduhudu
Morwamotsu
Motokwe
Morwamosi
Bulukubalo
Kokong
Mabutsane
Sekoma
Mahotshwane
Dutlwe
Takatokwane
Khudumel
Maboane
Tshwabi
Ditsheg
Mmakgodi
Jwaneng
Sese
Khakhea
Kakia
Tubani
Mokhomba
Tshono
Marushela
Lefoko
Lokerane
Sevrelela
Kokotsha
Khoti
Mabuasehube
Gasita
Machaakha
Kwedia
Lorolwana
Bray
Terra Firma
Sekhutlane
Khisa
Omaweneno
Lehwerane
Khanyeng
Tshabong
Vorstershoop
Pomfret
Setabeng
Tosca
Vergeleë
Tshidilamolomo
Gemsbokvlakte
Logageng
Tseoge
Kibitwe
Morokwaneng
Morokweng
Piet Plessis
Bona-Bona
Heuningvlei
Makhubung
Tsiloane
Perth
Khudunkgwane
Mosinki
Tlakgameng
Moswana
Ganyesa
Stella
Madibeng
Laxey
Loopeng
Madinonyane
Kgokgole
Southey
Louwna
Vryburg
Zylsrus
Metsimantsi
Wyk 3
Ditshipeng
Cassel
Bendel
Maphiniki
Tsineng
Maipeng
Dithakong
Hotazel
Sedibeng
Tswalu Game Reserve
Whyenbah
Winton
Mothibistad
Kamden
Kagung
Bothetheletsa
Ga-Masepa
Maganeng
Dry Harts
Pudimoe
Miles 0 50
Km 0 50
Map data ©2024 Google
©2024 F. Espenak
Map 05

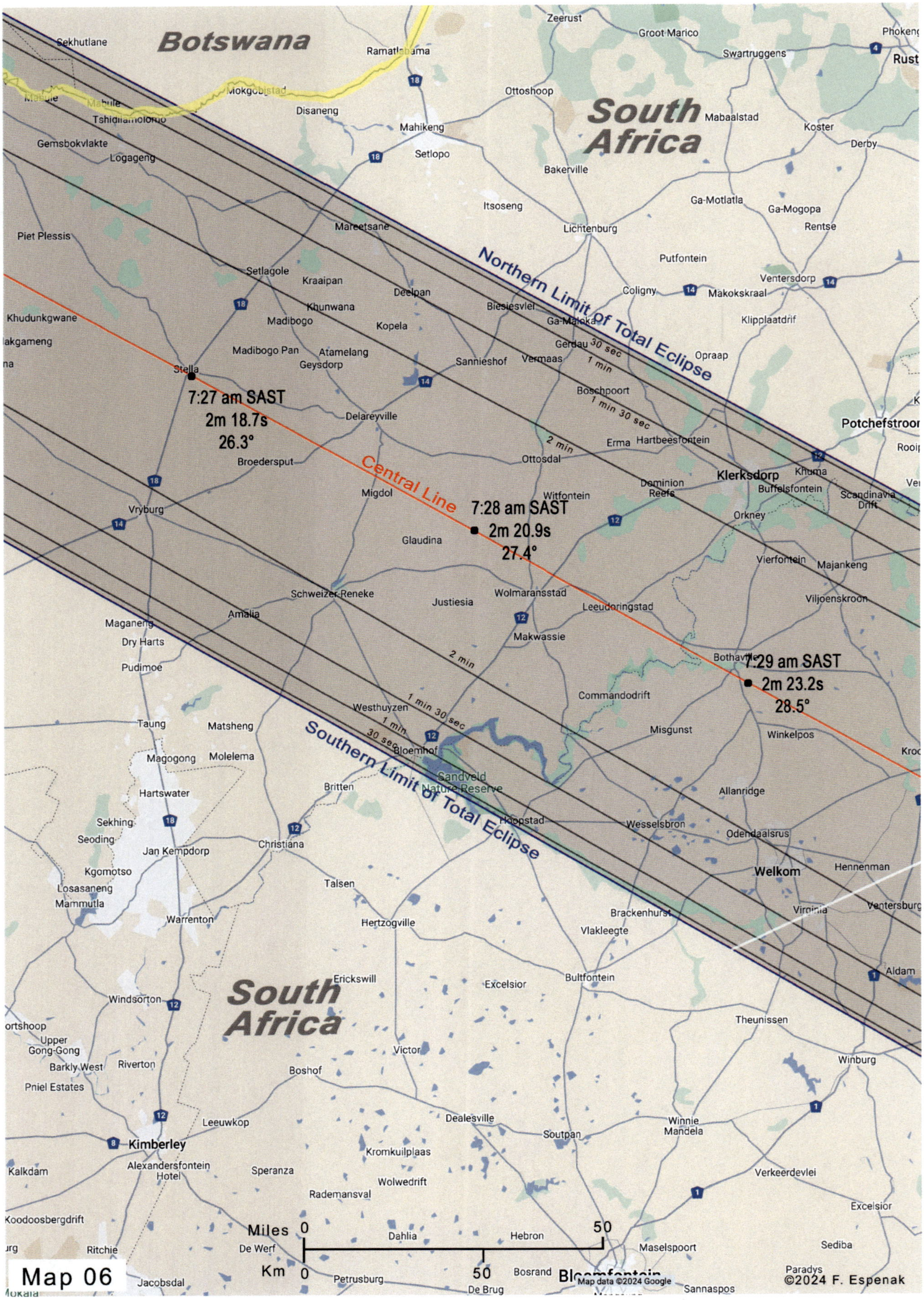
Botswana
South Africa
Northern Limit of Total Eclipse
Southern Limit of Total Eclipse
Central Line
30 sec
1 min
1 min 30 sec
2 min
7:27 am SAST
2m 18.7s
26.3°
7:28 am SAST
2m 20.9s
27.4°
7:29 am SAST
2m 23.2s
28.5°
Sekhutlane
Ramatlabama
Zeerust
Groot Marico
Swartruggens
Rust
Mokgobistad
Tshidilamolomo
Disaneng
Mahikeng
Ottoshoop
Mabaalstad
Koster
Derby
Gemsbokvlakte
Logageng
Setlopo
Bakerville
Itsoseng
Ga-Motlatla
Ga-Mogopa
Rentse
Piet Plessis
Mareetsane
Lichtenburg
Putfontein
Ventersdorp
Setlagole
Kraaipan
Deelpan
Coligny
Makokskraal
Khudunkgwane
Khunwana
Madibogo
Kopela
Biesiesvlei
Ga-Maloka
Klipplaatdrif
Gerdau
Madibogo Pan
Atamelang
Geysdorp
Sannieshof
Vermaas
Opraap
Stella
Boschpoort
Delareyville
Potchefstroom
Erma
Hartbeesfontein
Broedersput
Ottosdal
Klerksdorp
Khuma
Buffelsfontein
Dominion Reefs
Migdol
Witfontein
Scandinavia Drift
Vryburg
Orkney
Glaudina
Vierfontein
Majankeng
Schweizer-Reneke
Wolmaransstad
Justiesia
Viljoenskroon
Amalia
Leeudoringstad
Maganeng
Dry Harts
Makwassie
Bothaville
Pudimoe
Commandodrift
Westhuyzen
Taung
Matsheng
Misgunst
Winkelpos
Bloemhof
Magogong
Molelema
Sandveld Nature Reserve
Britten
Allanridge
Hartswater
Hoopstad
Wesselsbron
Sekhing
Seoding
Odendaalsrus
Christiana
Jan Kempdorp
Kgomotso
Losasaneng
Mammutla
Welkom
Hennenman
Talsen
Warrenton
Virginia
Ventersburg
Brackenhurst
Hertzogville
Vlakleegte
Aldam
Erickswill
Excelsior
Bultfontein
Windsorton
Theunissen
Upper Gong-Gong
Barkly West
Riverton
Victor
Winburg
Pniel Estates
Boshof
Leeuwkop
Dealesville
Winnie Mandela
Soutpan
Kimberley
Kromkuilplaas
Alexandersfontein Hotel
Speranza
Kalkdam
Wolwedrift
Verkeerdevlei
Rademansval
Koodoosbergdrift
Excelsior
Miles
0
50
Dahlia
Hebron
De Werf
Ritchie
Maselspoort
Sediba
Km
Petrusburg
Bosrand
De Brug
Jacobsdal
Sannaspos
Paradys
Map 06
Map data ©2024 Google
©2024 F. Espenak

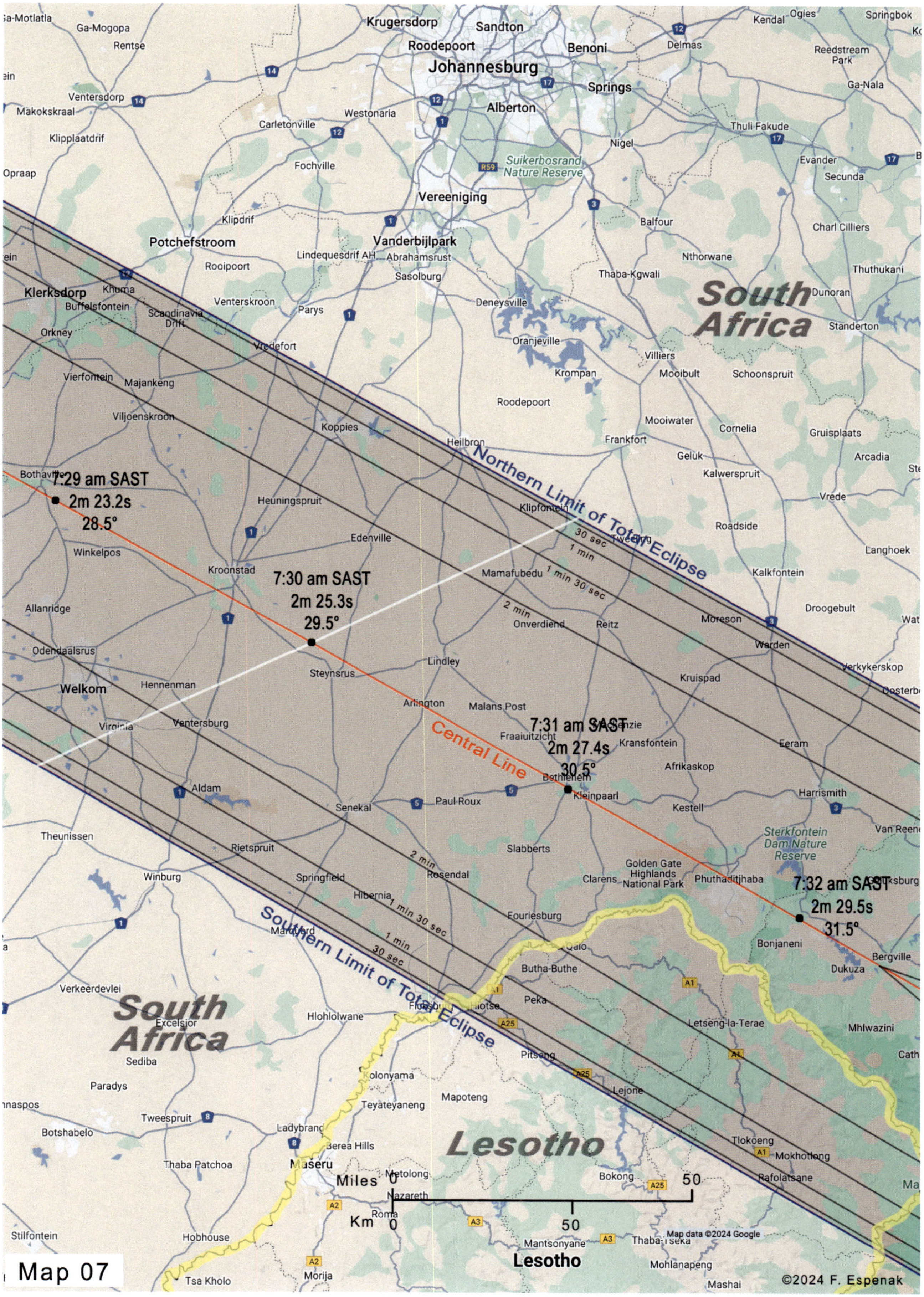
Map 07
7:29 am SAST
2m 23.2s
28.5°
7:30 am SAST
2m 25.3s
29.5°
7:31 am SAST
2m 27.4s
30.5°
7:32 am SAST
2m 29.5s
31.5°
Central Line
Northern Limit of Total Eclipse
Southern Limit of Total Eclipse
30 sec
1 min
1 min 30 sec
2 min
South Africa
Lesotho
Johannesburg
Krugersdorp
Sandton
Roodepoort
Benoni
Springs
Alberton
Westonaria
Carletonville
Fochville
Suikerbosrand Nature Reserve
Vereeniging
Vanderbijlpark
Sasolburg
Potchefstroom
Klerksdorp
Orkney
Parys
Vredefort
Heilbron
Frankfort
Koppies
Kroonstad
Edenville
Steynsrus
Lindley
Reitz
Bethlehem
Kleinpaarl
Senekal
Paul Roux
Welkom
Virginia
Ventersburg
Odendaalsrus
Allanridge
Winburg
Theunissen
Harrismith
Kestell
Clarens
Golden Gate Highlands National Park
Phuthaditjhaba
Fouriesburg
Butha-Buthe
Maseru
Bergville
Sterkfontein Dam Nature Reserve
Miles 0 50
Km 0 50
Map data ©2024 Google
©2024 F. Espenak

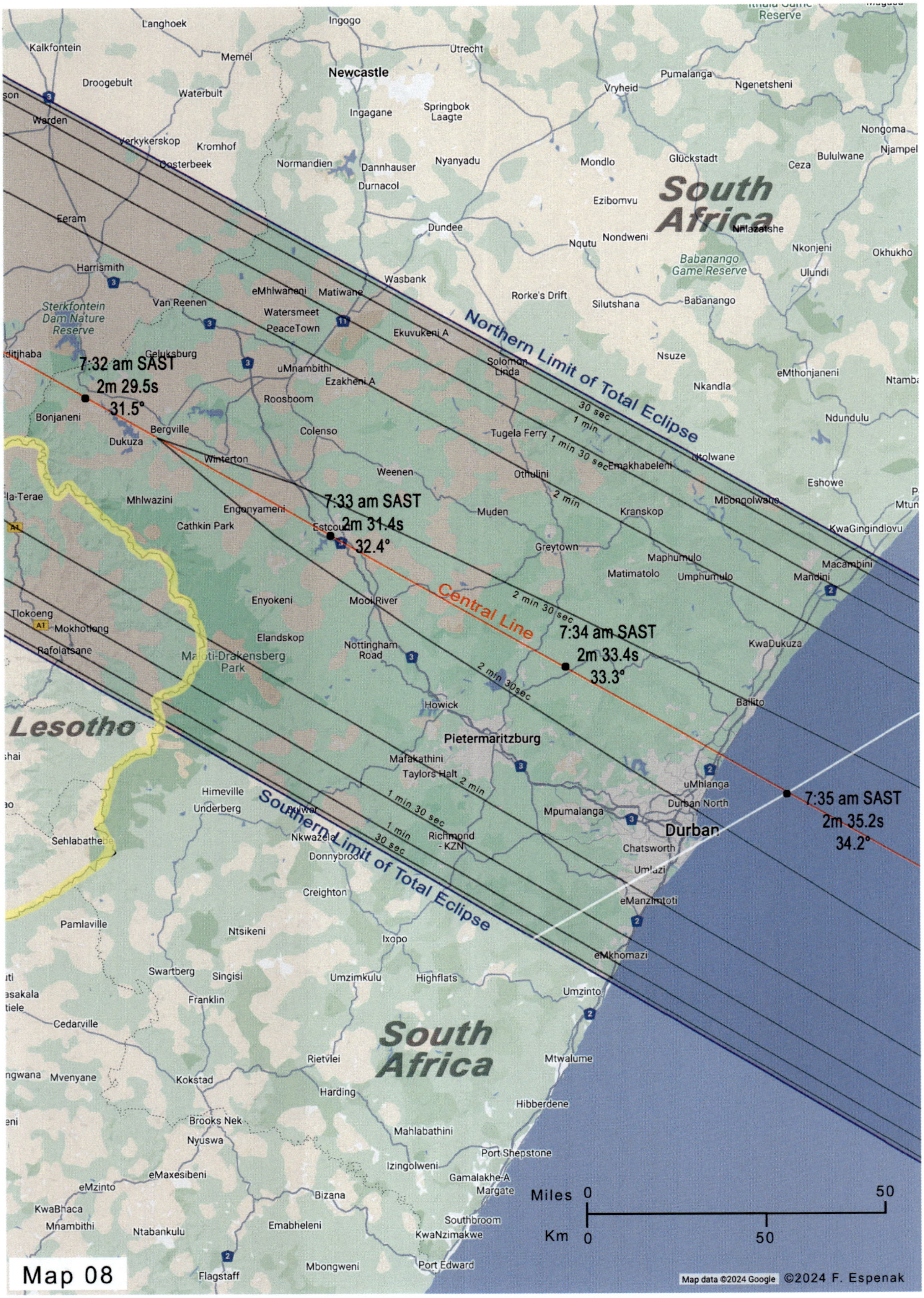
Map 08
Northern Limit of Total Eclipse
Southern Limit of Total Eclipse
Central Line
7:32 am SAST
2m 29.5s
31.5°
7:33 am SAST
2m 31.4s
32.4°
7:34 am SAST
2m 33.4s
33.3°
7:35 am SAST
2m 35.2s
34.2°
30 sec
1 min
1 min 30 sec
2 min
2 min 30 sec
2 min 30sec
2 min
1 min 30 sec
1 min
30 sec
South Africa
South Africa
Lesotho
Miles 0 50
Km 0 50
Map data ©2024 Google
©2024 F. Espenak
Newcastle
Pietermaritzburg
Durban
Langhoek
Ingogo
Kalkfontein
Memel
Utrecht
Droogebult
Waterbult
Vryheid
Pumalanga
Ngenetsheni
Warden
Ingagane
Springbok Laagte
Kromhof
Normandien
Dannhauser
Nyanyadu
Mondlo
Glückstadt
Nongoma
Bululwane
Ceza
Durnacol
Ezibomvu
Dundee
Nqutu
Nondweni
Eeram
Harrismith
Babanango Game Reserve
Nkonjeni
Ulundi
Okhukho
Wasbank
Rorke's Drift
Silutshana
Babanango
Van Reenen
Sterkfontein Dam Nature Reserve
eMhlwaneni
Matiwane
Watersmeet
PeaceTown
Ekuvukeni A
Solomon Linda
Nsuze
eMthonjaneni
Nkandla
uMnambithi
Ezakheni A
Roosboom
Bonjaneni
Bergville
Dukuza
Colenso
Tugela Ferry
Ndundulu
Winterton
Emakhabeleni
Weenen
Othulini
Eshowe
Mhlwazini
Engonyameni
Cathkin Park
Muden
Kranskop
KwaGingindlovu
Greytown
Maphumulo
Matimatolo
Umphumulo
Mandini
Enyokeni
Mooi River
Tlokoeng
Mokhotlong
Elandskop
Nottingham Road
KwaDukuza
Maloti-Drakensberg Park
Howick
Ballito
Taylors Halt
uMhlanga
Durban North
Himeville
Underberg
Mpumalanga
Sehlabathebe
Nkwazela
Donnybrook
Richmond - KZN
Chatsworth
Umlazi
eManzimtoti
Creighton
Pamlaville
Ntsikeni
Ixopo
eMkhomazi
Swartberg
Singisi
Umzimkulu
Highflats
Umzinto
Franklin
Cedarville
Rietvlei
Mtwalume
Mvenyane
Kokstad
Harding
Hibberdene
Brooks Nek
Nyuswa
Mahlabathini
Port Shepstone
Izingolweni
Gamalakhe-A
Margate
eMaxesibeni
eMzinto
KwaBhaca
Mnambithi
Bizana
Ntabankulu
Emabheleni
Southbroom
KwaNzimakwe
Port Edward
Mbongweni
Flagstaff

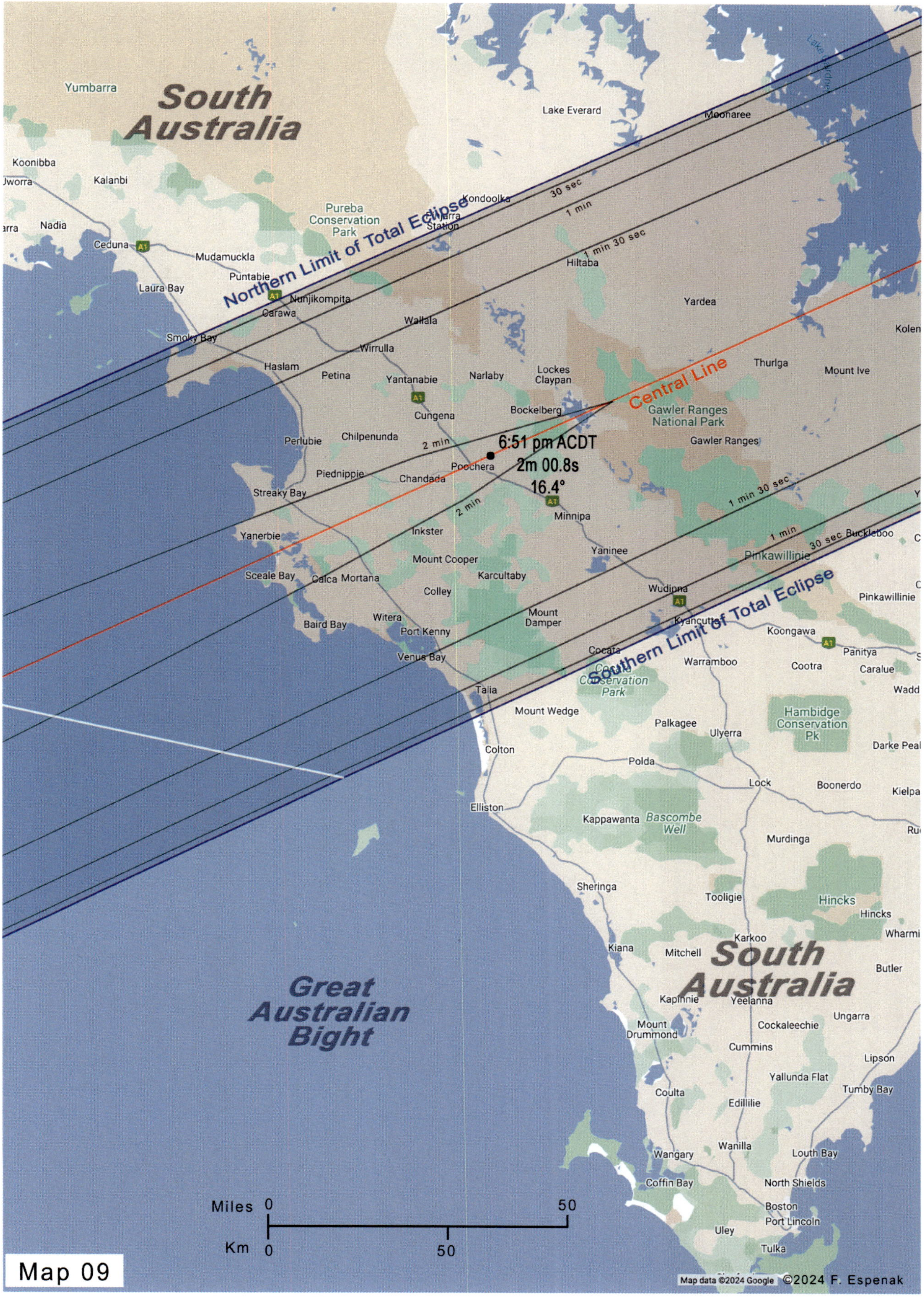
South Australia
Yumbarra
Koonibba
Kalanbi
Nadia
Ceduna
Pureba Conservation Park
Kondoolka
Yarra Station
Lake Everard
Moonaree
Mudamuckla
Laura Bay
Puntabie
Nunjikompita
Carawa
Smoky Bay
Northern Limit of Total Eclipse
30 sec
1 min
1 min 30 sec
Hiltaba
Yardea
Wallala
Wirrulla
Haslam
Petina
Yantanabie
Narlaby
Lockes Claypan
Thurlga
Mount Ive
Kolen
Central Line
Bockelberg
Cungena
Gawler Ranges National Park
Gawler Ranges
Perlubie
Chilpenunda
2 min
6:51 pm ACDT
2m 00.8s
16.4°
Poochera
Piednippie
Chandada
Streaky Bay
Minnipa
Inkster
Yanerbie
Yaninee
Mount Cooper
Sceale Bay
Calca
Mortana
Karcultaby
Colley
Wudinna
Pinkawillinie
Buckleboo
Baird Bay
Witera
Mount Damper
Port Kenny
Kyancutta
Southern Limit of Total Eclipse
Koongawa
Venus Bay
Cocata
Cocata Conservation Park
Warramboo
Cootra
Panitya
Caralue
Talia
Mount Wedge
Palkagee
Ulyerra
Hambidge Conservation Pk
Colton
Polda
Darke Peak
Lock
Boonerdo
Kielpa
Elliston
Kappawanta
Bascombe Well
Murdinga
Sheringa
Tooligie
Hincks
Wharminda
Kiana
Mitchell
Karkoo
Butler
Great Australian Bight
Kapinnie
Yeelanna
Mount Drummond
Cockaleechie
Ungarra
Cummins
Lipson
Yallunda Flat
Coulta
Tumby Bay
Edillilie
Wangary
Wanilla
Louth Bay
Coffin Bay
North Shields
Boston
Port Lincoln
Uley
Tulka
Miles 0 50
Km 0 50
Map data ©2024 Google
©2024 F. Espenak

Map 09

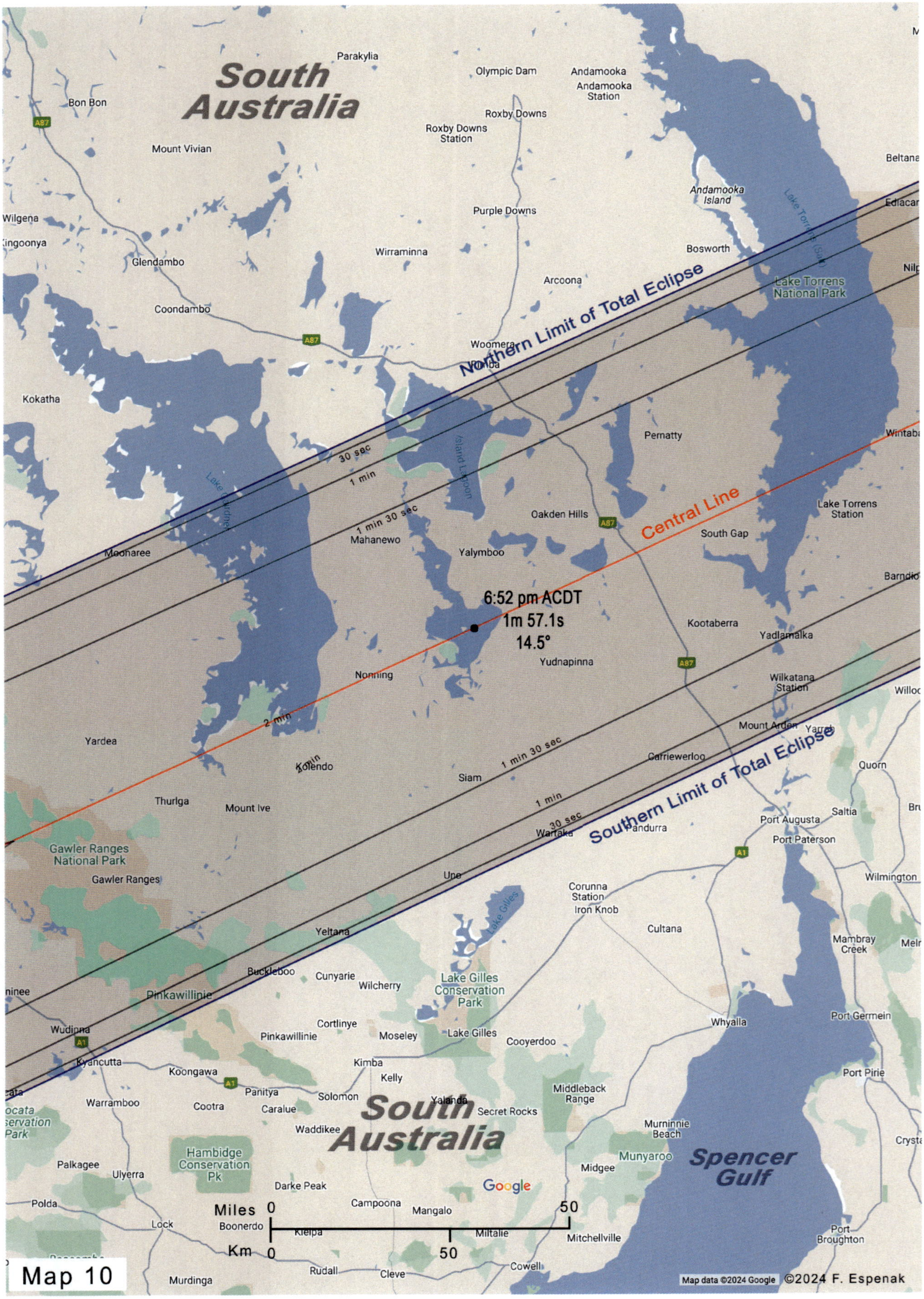
South Australia
Parakylia
Olympic Dam
Andamooka
Andamooka Station
Roxby Downs
Roxby Downs Station
Bon Bon
Mount Vivian
Beltana
Andamooka Island
Lake Torrens (Salt)
Wilgena
Kingoonya
Purple Downs
Bosworth
Glendambo
Wirraminna
Arcoona
Lake Torrens National Park
Coondambo
Northern Limit of Total Eclipse
Woomera
Pimba
Kokatha
Pernatty
Island Lagoon
Lake Gairdner
30 sec
1 min
1 min 30 sec
Oakden Hills
Lake Torrens Station
Central Line
South Gap
Mahanewo
Moonaree
Yalymboo
6:52 pm ACDT
1m 57.1s
14.5°
Kootaberra
Yadlamalka
Yudnapinna
Nonning
Wilkatana Station
2 min
Yardea
Mount Arden
Yarrah
Kolendo
Siam
1 min 30 sec
Carriewerloo
Southern Limit of Total Eclipse
Quorn
Thurlga
Mount Ive
1 min
30 sec
Wartaka
Pandurra
Port Augusta
Saltia
Port Paterson
Gawler Ranges National Park
Gawler Ranges
Uno
Corunna Station
Iron Knob
Wilmington
Cultana
Lake Gilles
Yeltana
Mambray Creek
Buckleboo
Cunyarie
Wilcherry
Lake Gilles Conservation Park
Pinkawillinie
Port Germein
Whyalla
Wudinna
Cortlinye
Pinkawillinie
Moseley
Lake Gilles
Cooyerdoo
Kyancutta
Koongawa
Kimba
Kelly
Port Pirie
Panitya
Solomon
Yalanda
Middleback Range
Warramboo
Cootra
Caralue
Secret Rocks
South Australia
Waddikee
Murninnie Beach
Hambidge Conservation Pk
Munyaroo
Spencer Gulf
Palkagee
Ulyerra
Midgee
Darke Peak
Google
Polda
Campoona
Mangalo
Miles 0
50
Lock
Boonerdo
Kielpa
Miltalie
Mitchellville
Port Broughton
Km 0
50
Map 10
Rudall
Cleve
Cowell
Murdinga
Map data ©2024 Google
©2024 F. Espenak

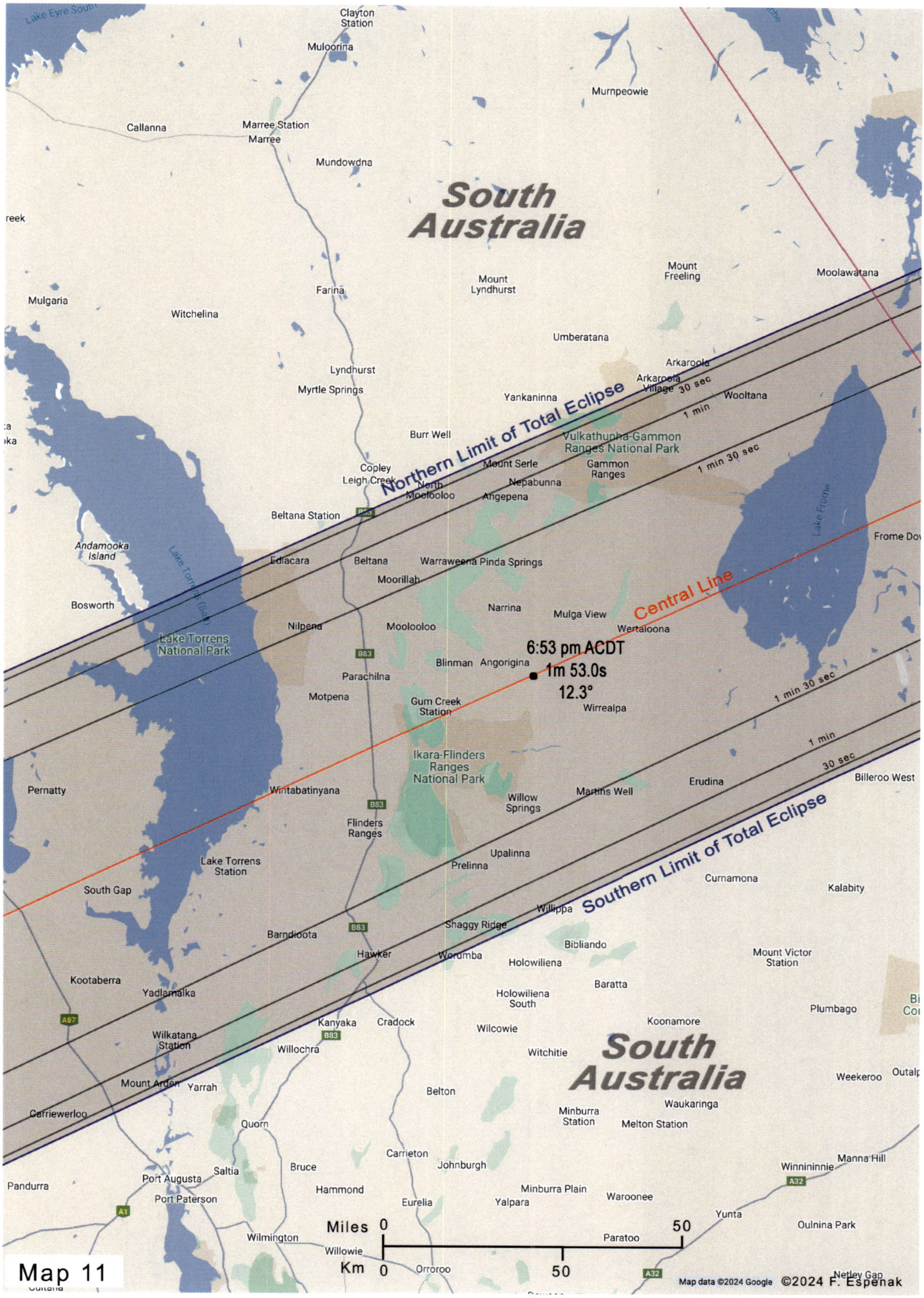
South Australia
Northern Limit of Total Eclipse
Central Line
Southern Limit of Total Eclipse
30 sec
1 min
1 min 30 sec
6:53 pm ACDT
1m 53.0s
12.3°
Clayton Station
Muloorina
Murnpeowie
Callanna
Marree Station
Marree
Mundowdna
Mulgaria
Witchelina
Farina
Mount Lyndhurst
Mount Freeling
Moolawatana
Umberatana
Lyndhurst
Myrtle Springs
Arkaroola
Arkaroola Village
Yankaninna
Wooltana
Burr Well
Vulkathunha-Gammon Ranges National Park
Copley
Leigh Creek
North Moolooloo
Mount Serle
Nepabunna
Gammon Ranges
Angepena
Beltana Station
Andamooka Island
Ediacara
Beltana
Warraweena
Pinda Springs
Moorillah
Frome Downs
Bosworth
Nilpena
Moolooloo
Narrina
Mulga View
Wertaloona
Lake Torrens National Park
Blinman
Angorigina
Parachilna
Motpena
Gum Creek Station
Wirrealpa
Ikara-Flinders Ranges National Park
Pernatty
Wintabatinyana
Willow Springs
Martins Well
Erudina
Billeroo West
Flinders Ranges
Lake Torrens Station
Upalinna
Prelinna
South Gap
Curnamona
Kalabity
Willippa
Barndioota
Shaggy Ridge
Hawker
Worumba
Bibliando
Holowiliena
Mount Victor Station
Kootaberra
Yadlamalka
Baratta
Holowiliena South
Plumbago
Kanyaka
Cradock
Koonamore
Wilkatana Station
Wilcowie
Willochra
Witchitie
Mount Arden
Yarrah
Weekeroo
Belton
Carriewerloo
Waukaringa
Minburra Station
Melton Station
Quorn
Carrieton
Manna Hill
Winnininnie
Saltia
Bruce
Johnburgh
Pandurra
Port Augusta
Hammond
Minburra Plain
Waroonee
Port Paterson
Eurelia
Yalpara
Yunta
Oulnina Park
Wilmington
Paratoo
Willowie
Orroroo
Miles 0 50
Km 0 50
Map 11
Map data ©2024 Google
©2024 F. Espenak
Netley Gap

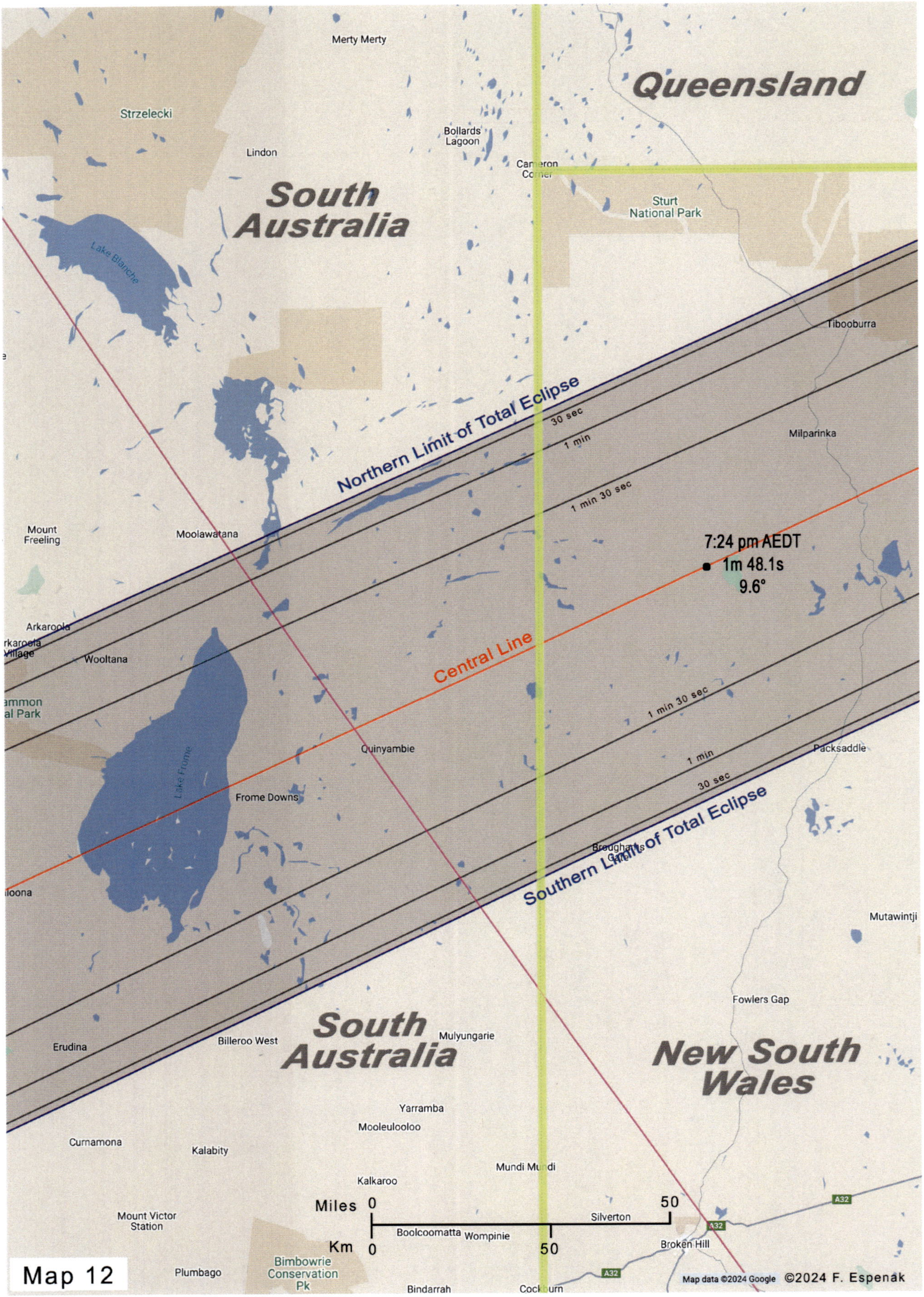
Queensland
South Australia
New South Wales
South Australia
Merty Merty
Strzelecki
Lindon
Bollards Lagoon
Cameron Corner
Sturt National Park
Lake Blanche
Tibooburra
Milparinka
Northern Limit of Total Eclipse
30 sec
1 min
1 min 30 sec
Mount Freeling
Moolawatana
7:24 pm AEDT
1m 48.1s
9.6°
Arkaroola
Wooltana
Central Line
1 min 30 sec
Quinyambie
Packsaddle
1 min
30 sec
Lake Frome
Frome Downs
Southern Limit of Total Eclipse
Mutawintji
Fowlers Gap
Erudina
Billeroo West
Mulyungarie
Yarramba
Mooleulooloo
Curnamona
Kalabity
Mundi Mundi
Kalkaroo
Miles 0 50
Km 0 50
Mount Victor Station
Boolcoomatta
Wompinie
Silverton
Broken Hill
A32
Plumbago
Bimbowrie Conservation Pk
Bindarrah
Map 12
Map data ©2024 Google
©2024 F. Espenak

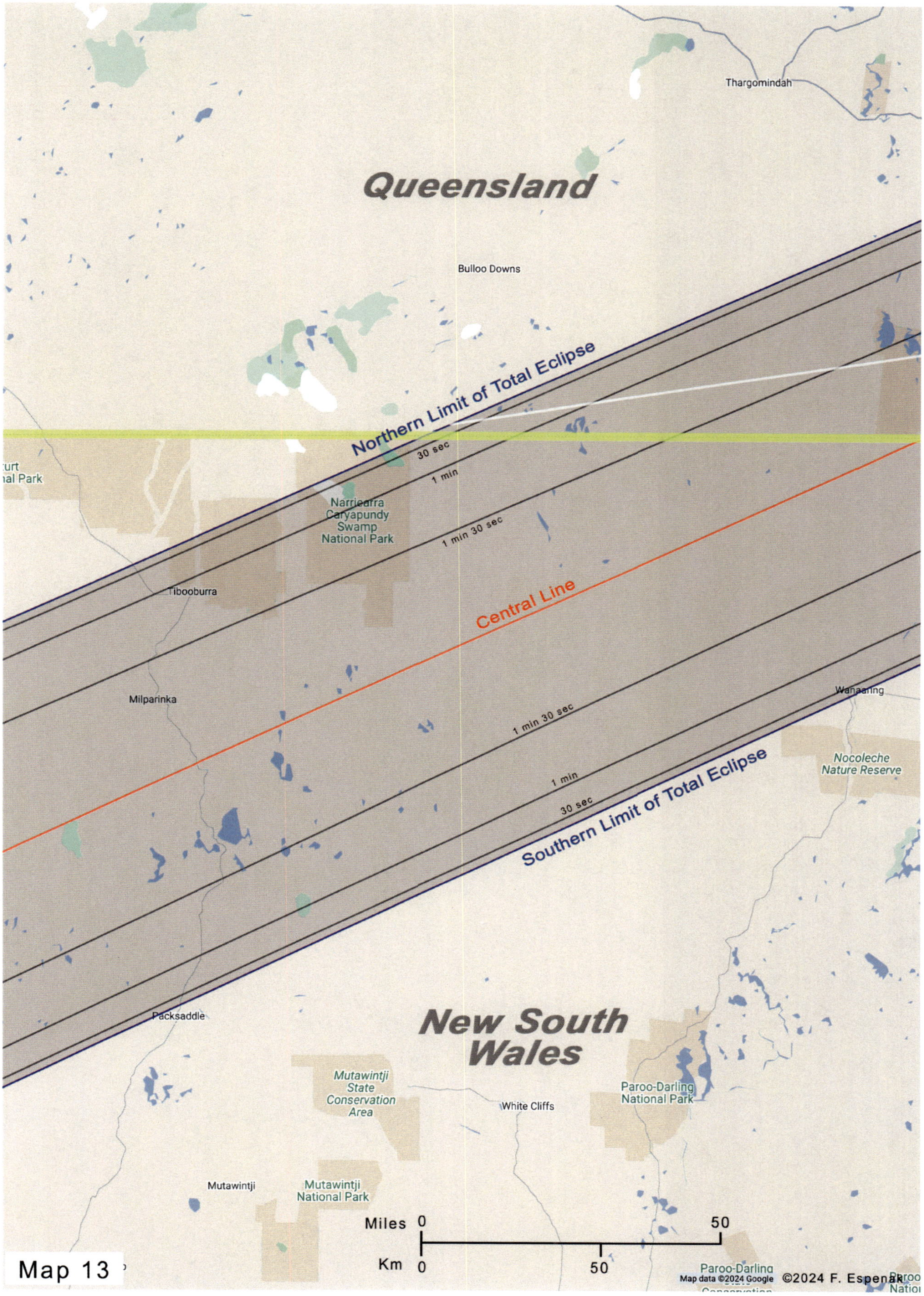
Thargomindah
Queensland
Bulloo Downs
Northern Limit of Total Eclipse
30 sec
1 min
1 min 30 sec
Narriearra Caryapundy Swamp National Park
Tibooburra
Central Line
Milparinka
Wanaaring
1 min 30 sec
Nocoleche Nature Reserve
1 min
30 sec
Southern Limit of Total Eclipse
Packsaddle
New South Wales
Mutawintji State Conservation Area
White Cliffs
Paroo-Darling National Park
Mutawintji
Mutawintji National Park
Miles 0 50
Km 0 50
Map 13
Paroo-Darling
Map data ©2024 Google
©2024 F. Espenak

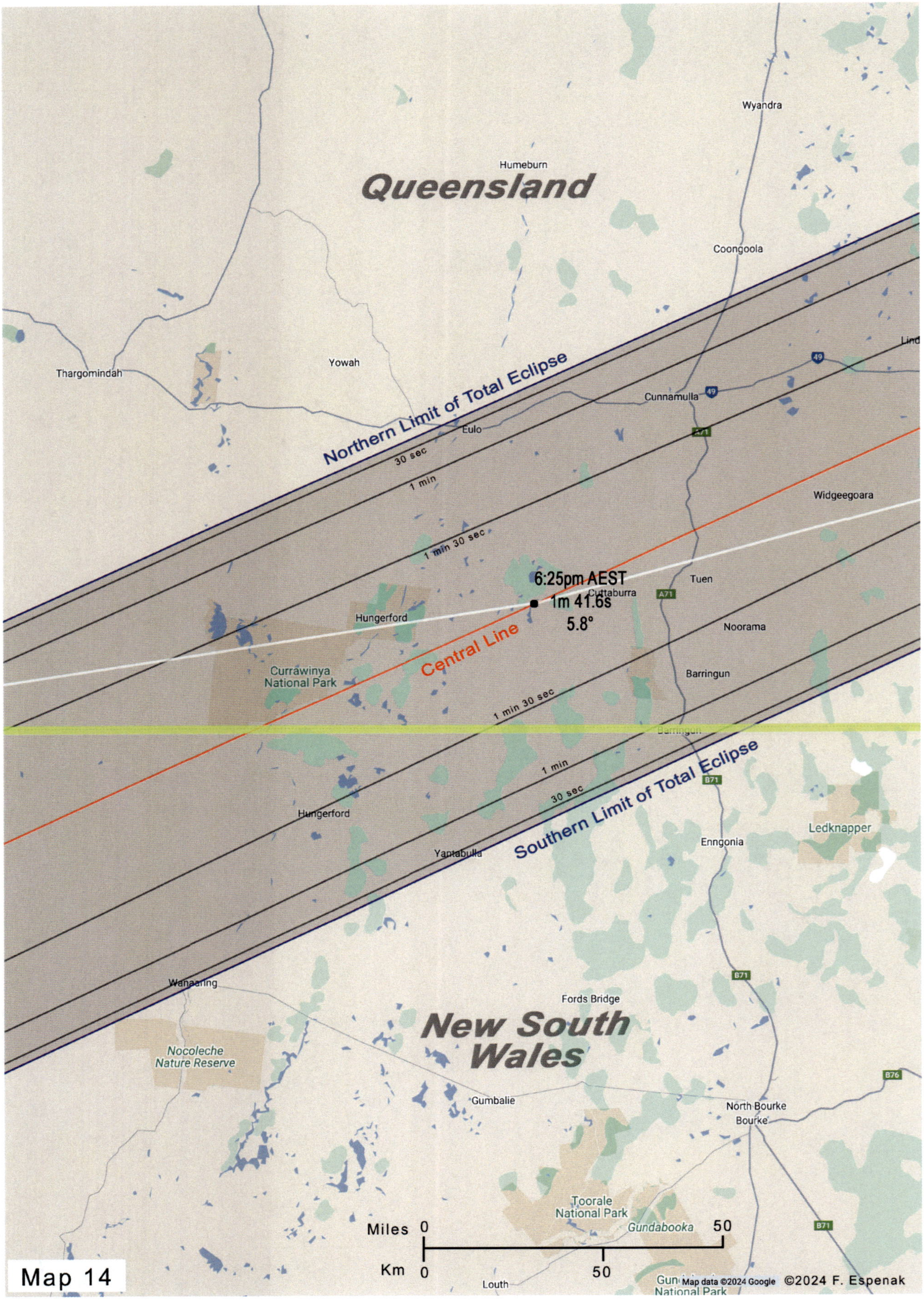

Queensland
Humeburn
Wyandra
Coongoola
Thargomindah
Yowah
Cunnamulla
Eulo
Northern Limit of Total Eclipse
30 sec
1 min
1 min 30 sec
Widgeegoara
6:25pm AEST
1m 41.6s
5.8°
Cuttaburra
Tuen
Hungerford
Noorama
Currawinya National Park
Central Line
Barringun
1 min 30 sec
1 min
30 sec
Southern Limit of Total Eclipse
Hungerford
Yantabulla
Enngonia
Ledknapper
Wanaaring
Fords Bridge
New South Wales
Nocoleche Nature Reserve
Gumbalie
North Bourke
Bourke
Toorale National Park
Gundabooka
Miles 0 50
Km 0 50
Louth
Map 14
Map data ©2024 Google
©2024 F. Espenak

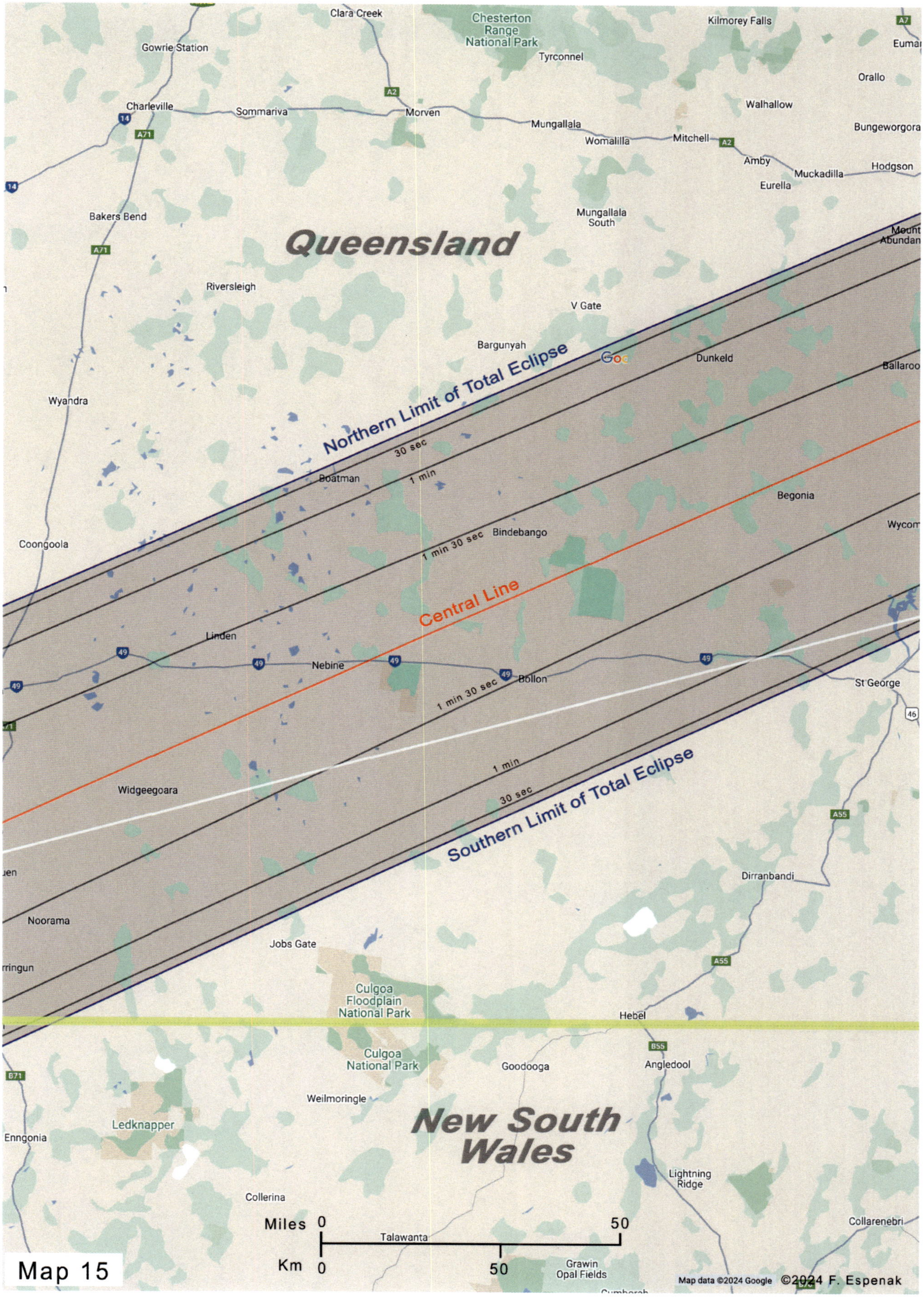
Clara Creek
Chesterton Range National Park
Kilmorey Falls
Gowrie Station
Tyrconnel
Orallo
Charleville
Sommariva
Morven
Mungallala
Womalilla
Mitchell
Walhallow
Bungeworgora
Amby
Muckadilla
Hodgson
Eurella
Bakers Bend
Mungallala South
Queensland
Riversleigh
V Gate
Bargunyah
Dunkeld
Ballaroo
Wyandra
Northern Limit of Total Eclipse
30 sec
1 min
Boatman
Begonia
Coongoola
1 min 30 sec
Bindebango
Central Line
Linden
Nebine
Bollon
St George
Widgeegoara
Southern Limit of Total Eclipse
Dirranbandi
Noorama
Jobs Gate
Culgoa Floodplain National Park
Hebel
Culgoa National Park
Goodooga
Angledool
Weilmoringle
Ledknapper
Enngonia
New South Wales
Lightning Ridge
Collerina
Collarenebri
Miles 0 50
Talawanta
Km 0 50
Grawin Opal Fields
Map 15
Map data ©2024 Google
©2024 F. Espenak

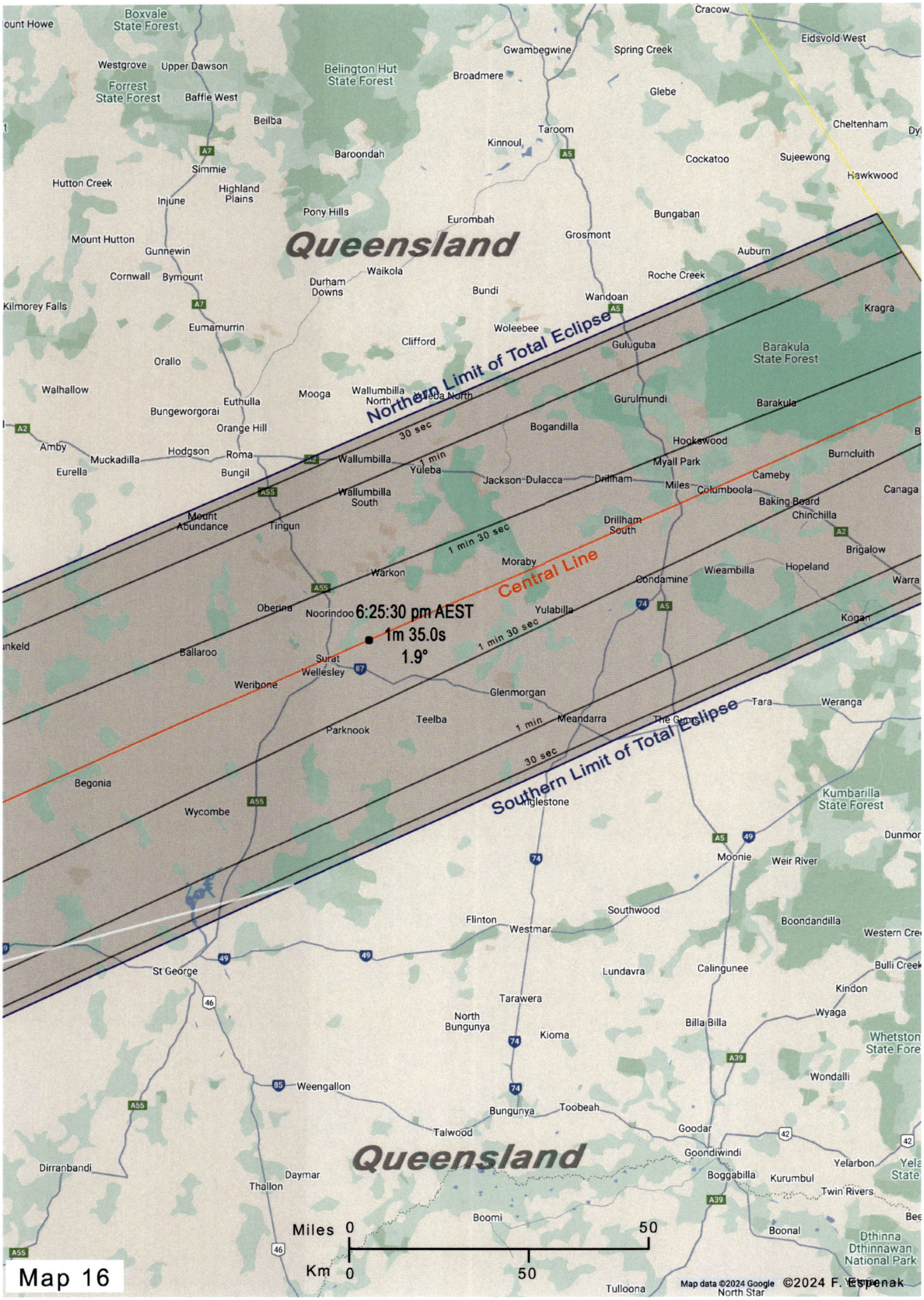

Queensland
Northern Limit of Total Eclipse
Southern Limit of Total Eclipse
Central Line
30 sec
1 min
1 min 30 sec
6:25:30 pm AEST
1m 35.0s
1.9°
Roma
Surat
St George
Miles
Chinchilla
Condamine
Moonie
Goondiwindi
Taroom
Wandoan
Tara
Miles 0 50
Km 0 50
Map 16
Map data ©2024 Google
©2024 F. Espenak

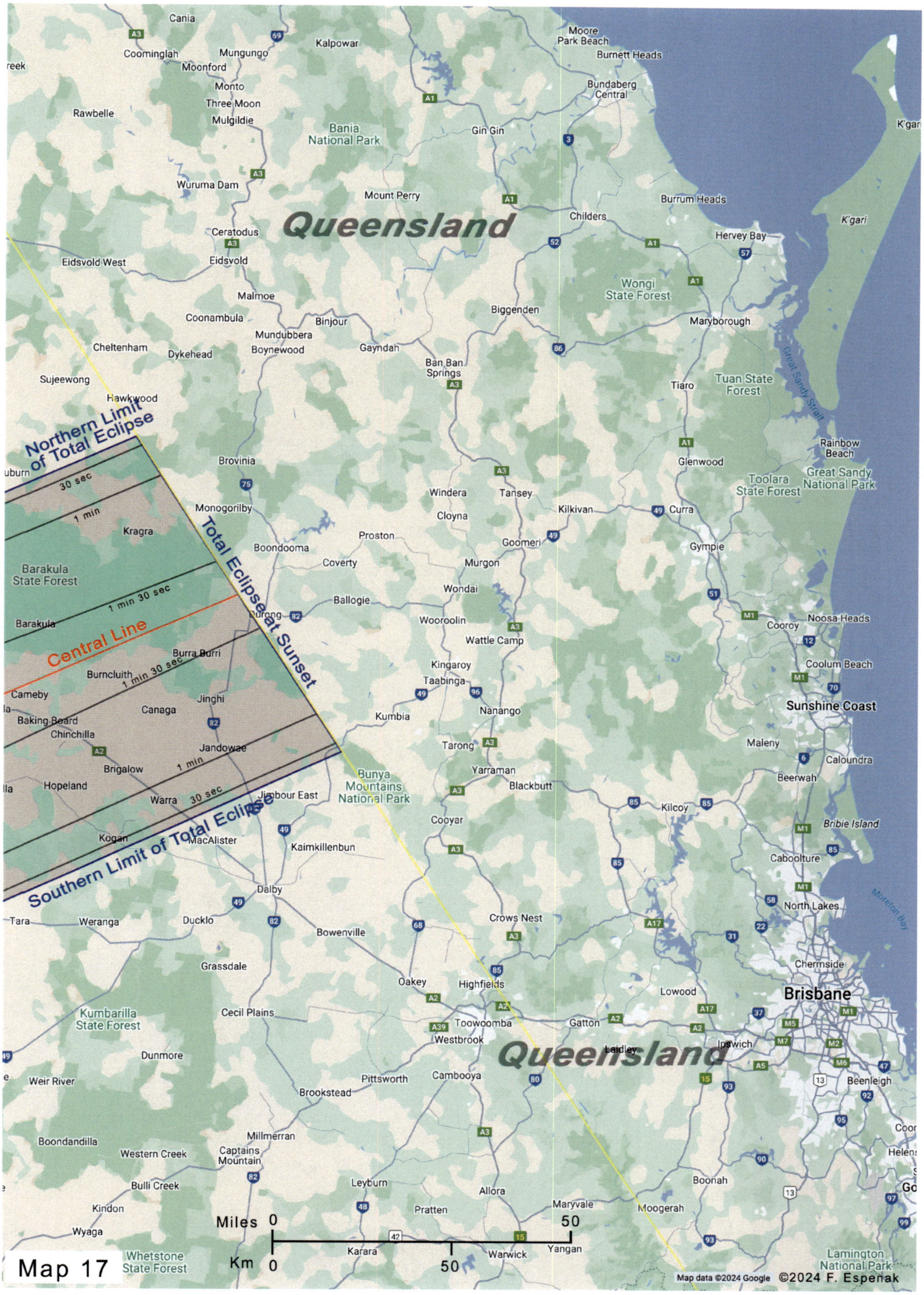
Queensland
Northern Limit of Total Eclipse
Southern Limit of Total Eclipse
Central Line
Total Eclipse at Sunset
30 sec
1 min
1 min 30 sec
1 min 30 sec
1 min
30 sec
Cania
Coominglah
Mungungo
Kalpowar
Moonford
Monto
Three Moon
Mulgildie
Rawbelle
Bania National Park
Moore Park Beach
Burnett Heads
Bundaberg Central
Gin Gin
Wuruma Dam
Mount Perry
Burrum Heads
Childers
Ceratodus
Eidsvold
Eidsvold West
Hervey Bay
K'gari
Wongi State Forest
Malmoe
Coonambula
Binjour
Biggenden
Mundubbera
Boynewood
Gayndah
Cheltenham
Dykehead
Maryborough
Ban Ban Springs
Sujeewong
Hawkwood
Tiaro
Tuan State Forest
Great Sandy Strait
Rainbow Beach
Glenwood
Brovinia
Toolara State Forest
Great Sandy National Park
Windera
Tansey
Monogorilby
Kilkivan
Curra
Cloyna
Kragra
Proston
Goomeri
Gympie
Boondooma
Coverty
Murgon
Barakula State Forest
Wondai
Ballogie
Durong
Barakula
Wooroolin
Cooroy
Noosa Heads
Wattle Camp
Burra Burri
Kingaroy
Coolum Beach
Taabinga
Burncluith
Cameby
Jinghi
Nanango
Sunshine Coast
Kumbia
Canaga
Baking Board
Chinchilla
Tarong
Maleny
Jandowae
Caloundra
Brigalow
Yarraman
Beerwah
Bunya Mountains National Park
Blackbutt
Hopeland
Jimbour East
Warra
Kilcoy
Bribie Island
Cooyar
Kogan
MacAlister
Kaimkillenbun
Caboolture
Dalby
North Lakes
Moreton Bay
Tara
Weranga
Ducklo
Crows Nest
Bowenville
Chermside
Grassdale
Oakey
Highfields
Lowood
Brisbane
Kumbarilla State Forest
Cecil Plains
Toowoomba
Gatton
Westbrook
Laidley
Ipswich
Dunmore
Weir River
Pittsworth
Cambooya
Brookstead
Beenleigh
Millmerran
Boondandilla
Western Creek
Captains Mountain
Bulli Creek
Leyburn
Allora
Boonah
Kindon
Pratten
Maryvale
Moogerah
Wyaga
Karara
Warwick
Yangan
Whetstone State Forest
Lamington National Park
Miles 0 50
Km 0 50
Map 17
Map data ©2024 Google
©2024 F. Espenak

Eclipse Guides from AstroPixels Publishing

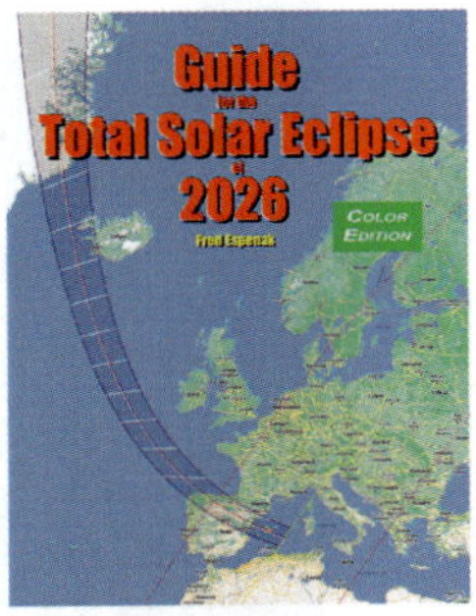

Guide for the Total Solar Eclipse of 2026

The total solar eclipse of August 12, 2026 will be visible from a path across northern Russia, the Arctic, Greenland, Iceland, and Spain. The **Guide for the Total Solar Eclipse of 2026** contains 14 full-page maps of the path. The large scale shows roads, towns and cities, rivers and lakes.

Tables for hundreds of cities provide times each eclipse phase of the along with the eclipse magnitude, duration, and the Sun's altitude. A climatology study identifies areas along the eclipse path where the highest probability of favorable weather may be found.

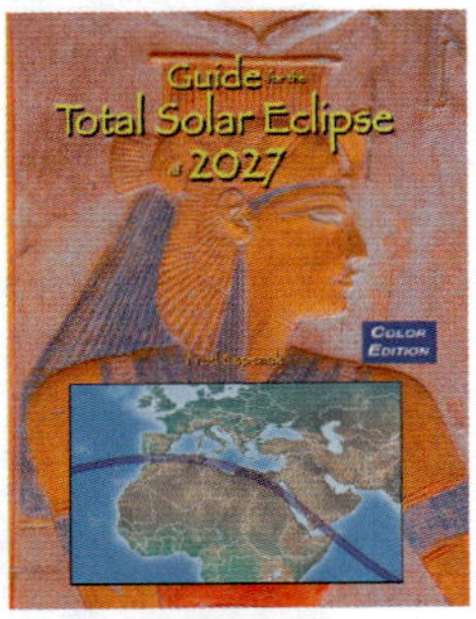

Guide for the Total Solar Eclipse of 2027

The total solar eclipse of August 2, 2027 will be visible from a path across Spain, Morocco, Algeria, Tunisia, Libya, Egypt, Sudan, Saudia Arabia, Yemen, and Somalia. The **Guide for the Total Solar Eclipse of 2027** contains 20 full-page maps of the path.

Tables for hundreds of cities provide times for each eclipse phase along with the eclipse magnitude, duration and the Sun's altitude. Cloud statistics identify areas where the highest probability of clear sky may be found.

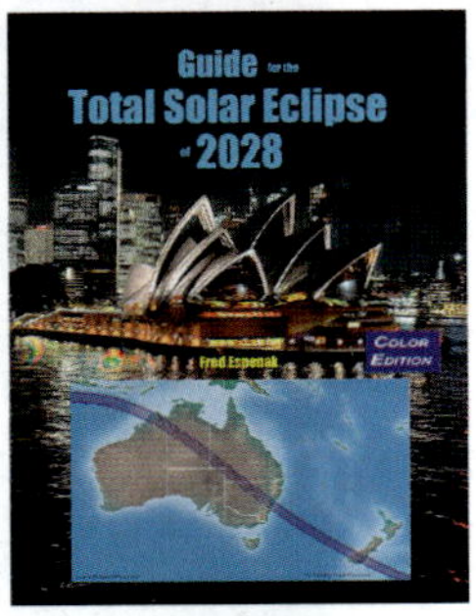

Guide for the Total Solar Eclipse of 2028

The total solar eclipse of July 22, 2028 will be visible from a path across Australia and New Zealand. The **Guide for the Total Solar Eclipse of 2028** contains 19 full-page maps of the path of totality.

Tables for over one hundred cities provide times for each eclipse phase with the duration of totality and the Sun's altitude. Cloud statistics identify areas where the highest probability of clear sky may be found. A preview is given for total eclipses through 2039.

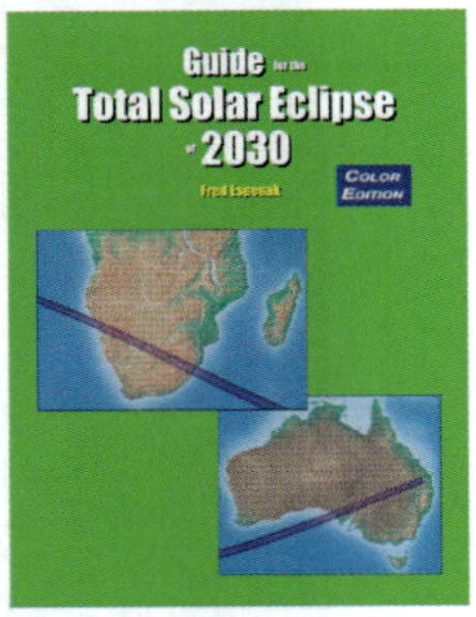

Guide for the Total Solar Eclipse of 2030

The total solar eclipse of November 25, 2030 will be visible from a path across Africa and Australia. The **Guide for the Total Solar Eclipse of 2028** contains 17 full-page maps of the path of totality.

Tables for over one hundred cities provide times for each eclipse phase with the duration of totality and the Sun's altitude. Cloud statistics identify areas where the highest probability of clear sky may be found. A preview is given for total eclipses through 2039.

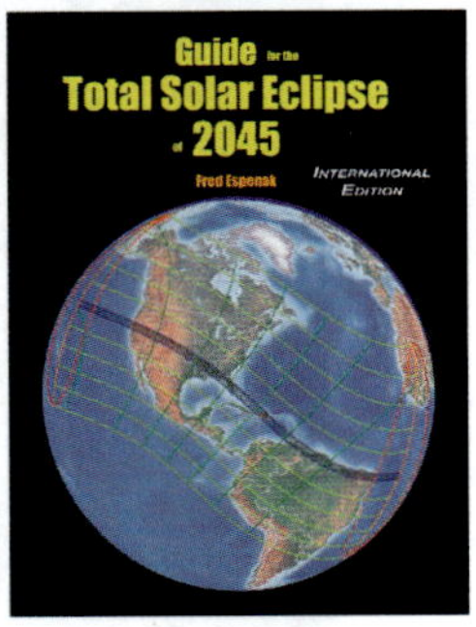

Guide for the Total Solar Eclipse of 2045

The total solar eclipse of August 12, 2045 will be visible from a path across the USA, the Caribbean, and South America. The **Guide for the Total Solar Eclipse of 2045** contains 37 full-page maps of the path of totality.

Tables for hundreds cities provide times for each eclipse phase with the duration of totality, eclipse magnitude, and the Sun's altitude. Cloud statistics identify areas where the highest probability of clear sky may be found.

For more information on each *Guide*, including sample pages of each, visit:

astropixels.com/pubs/guides.html

The *Guides* may be ordered from ***Amazon***

Eclipse Books from AstroPixels Publishing

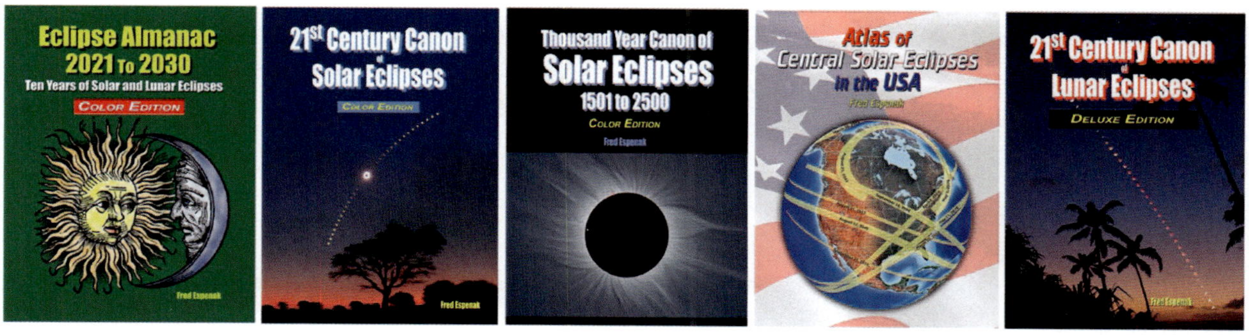

Eclipse Almanac – 2021 to 2030

A concise reference for every eclipse of the Sun and the Moon over a 10-year period. It identifies when and where each event will be seen with maps for each. Additional details about each eclipse are included. Other volumes in this series include the decades 2032 to 2040, 2041 to 2050, 2051 to 2060, and 2061 to 2070.

21st Century Canon of Solar Eclipses

The complete guide to every solar eclipse occurring from 2001 tom 2100 (224 eclipses in all). It includes information and maps for all total, annular, hybrid, and partial eclipses. A special world atlas shows detailed, full page maps of all central eclipse paths (total, annular, and hybrid).

Thousand Year Canon of Solar Eclipses — 1501 to 2500

Contains maps and data for each of the 2,389 solar eclipses occurring over the ten-century period centered on the present era. The maps are arranged twelve to a page at an image scale allowing the determination of eclipse visibility from any location on Earth. A comprehensive catalog lists the essential characteristics of each eclipse.

Atlas of Central Solar Eclipses in the USA

When was the last total eclipse through the USA and when is the next? How often do they happen? What total eclipse tracks passed across the USA during the 17th, 18th, and 19th centuries, etc., and what states did they include? And how often is a total solar eclipse visible from each of the 50 states? The Atlas of Central Solar Eclipses in the USA answers all these questions and more, with hundreds of maps and tables.

21st Century Canon of Lunar Eclipses

The complete guide to every lunar eclipse occurring from 2001 tom 2100 (228 eclipses in all). It includes information and maps for all total, partial, and penumbral eclipses. The predictions use a new model for Earth's elliptical shadows.

Thousand Year Canon of Lunar Eclipses — 1501 to 2500

Contains maps and data for each of the 2,424 lunar eclipses occurring over the ten-century period centered on the present era. The maps are arranged twelve to a page at an image scale allowing the determination of eclipse visibility from any location on Earth. A comprehensive catalog lists the essential characteristics of each eclipse.

All books are available in two editions: 1) Black and White, and 2) Color.

For more information including sample pages of each, visit:

astropixels.com/pubs/guides.html

Each *Book* may be ordered from ***Amazon***

Made in the USA
Coppell, TX
27 September 2026

91150337R00026